Working with Political Science Research Methods

Problems and Exercises

Third Edition

Jason D. Mycoff
University of Delaware

⑤SAGE | **CQPRESS**

Los Angeles | London | New Delhi
Singapore | Washington DC

Los Angeles | London | New Delhi
Singapore | Washington DC

FOR INFORMATION:

CQ Press
An Imprint of SAGE Publications, Inc.
2455 Teller Road
Thousand Oaks, California 91320
E-mail: order@sagepub.com

SAGE Publications Ltd.
1 Oliver's Yard
55 City Road
London, EC1Y 1SP
United Kingdom

SAGE Publications India Pvt. Ltd.
B 1/I 1 Mohan Cooperative Industrial Area
Mathura Road, New Delhi 110 044
India

SAGE Publications Asia-Pacific Pte. Ltd.
33 Pekin Street #02-01
Far East Square
Singapore 048763

Printed in the United States of America

ISBN: 978-1-60871-690-6

This book is printed on acid-free paper.

12 13 14 15 10 9 8 7 6 5 4 3 2

Acquisitions Editor: Charisse Kiino
Production Editor: Sarah Fell
Copy Editor: Pam Suwinsky
Typesetter: C&M Digitals (P) Ltd.
Cover Designer: Kimberly Glyder Design
Marketing Manager: Chris O'Brien

Contents

CHAPTER 1

Introduction
Practice Makes Perfect

While most political science courses deal with government, issues, and politics, research methods is an important component of political science that helps students better understand course work and has practical applications. Most political arguments raise claims of fact, as when someone says, "We *should* repeal the Obama health plan because it will *increase* health spending." The first part of the statement takes a normative position (something ought to be done), whereas the second makes a factual claim; it states that one thing leads to another, whether or not anyone wants it to be the case. The goal of our textbook, *Political Science Research Methods,* is to show you how those two types of assertions can be separated and how the latter can be demonstrated empirically. The goal of this workbook is to help you apply those cognitive skills.

At first sight, achieving these objectives may seem easy. And it is! But it also requires a degree of thought and care. Moreover, the best way to acquire the necessary skills is to practice actively and then practice some more. After all, no sports team would prepare for a game simply by reading a scouting report. But we believe if you make an honest effort, the process of verification can be fascinating as well as informative.

Most of the exercises in this workbook ask you to think before writing. The thought process is typically straightforward and certainly does not require a strong mathematical aptitude. A thorough reading of the text and attention to class notes plus a dose of common sense should be adequate.

Note also that many questions call for judgment and explanation; they do not necessarily have one "correct" answer. Unless a question is based on a straightforward calculation, or reading of a table, you will often be asked to think about a possible solution and to defend your choice.

The chapters in this workbook follow the chapters in the main text. That is, there are exercises for each chapter except the first and last. It is important to read the chapter in the text *before* starting to do an assignment. Many questions require you to integrate a chapter's different elements. Hence, you cannot just try to look up something without grasping the subject matter as a whole.

An orderly, step-by-step approach is the best way to work through the exercises in this workbook; it will help you avoid errors and make the important concepts relayed in the text clearer. If you are asked to make any calculations, you should do them neatly on a separate piece of paper that you can, if the instructor approves, turn in along with your answers. Here's a tip: your intermediate calculations or scrap work should be written in such a way that someone can reconstruct your thought processes. Figure 1–1 provides a simple example. It shows that the respondent first clarified the requested information and then performed the computation on a separate sheet of paper.

Figure 1–2 gives an example of someone using the workbook itself as a scratch pad and the ensuing confusion that often comes from sloppy writing and thinking. Note that some of the numbers were copied incorrectly and that the arithmetical operations are out of order. (The correct answer, by the way, is $28,650, not $24,150.)

All the "data" you need to do the exercises in this workbook are included in the workbook or in the textbook or can be downloaded from the student Web site at http://psrm.cqpress.com. When you are looking for a specific data set on the student Web site, simply click on the appropriate folder on the Web site. For instance, if you are looking for anes2004.dat, anes2004.por, or anescodebook.txt, open the folder called "ANES Data," and you will find each of those files. Once you learn how to use a program it is easy to explore a variety of hypotheses and problems. Besides being intrinsically interesting, knowledge of research methods provides skills that will help you in other courses and in many professions.

Have fun!

FIGURE 1-1

Be Organized and Neat

Party	Freq.
Democrat	200
Republican	150
Independent	100
None	50
Total	500

[Separate sheet of paper]

Percent = Number over total times 100.

200 Dems/500 = 2/5 = .4

.4 × 100 = 40%

Refer to the table above.

What percentage of the sample are Democrats?   40%

FIGURE 1-2

Sloppiness Leads to Errors

What is the mean, or average, per capita income of the following six countries?   24,150

Luxembourg	$32,700
United States	31,500
Bermuda	30,000
Switzerland	26,400
Singapore	26,300
Hong Kong	25,000

$$\frac{144,900}{5 \quad 6} = 24,150$$

32,500 32,700

31,500

3,000

26,400

26,300

25,500 25,800 25,000

1,414,900

CHAPTER 2

The Empirical Approach to Political Science

In chapter 2 of the textbook we describe the scientific method and argue that it underlies empirical political science research. We note that empiricism is not the only method of obtaining knowledge—there are others that lots of people fervently adhere to—and a case can be made against trying to study politics scientifically. (There are even disagreements about the definition and nature of the scientific method.) Nevertheless, this way of acquiring knowledge is so common that many social scientists take it for granted. And so do many average citizens. The problem is that scientific claims are sometimes difficult to distinguish from other kinds of statements. Nor is it always clear whether and how empirical analysis can be applied to propositions stated in theoretical and practical terms. The following questions, problems, and assignments therefore offer opportunities for you to think about the application of the empirical approach. Note that not all the questions have one "right" answer. Many, in fact, require a lot of careful thought. And it is often necessary to redefine or clarify words or phrases, to look for hidden assumptions, and to consider whether or not statements can be "translated" into scientific terms.

Exercise 2–1. Make a list of the characteristics of scientific knowledge. The list may help organize your thinking for other questions in this chapter.

_____

_____

_____

_____

_____

_____

_____

_____

Exercise 2–2. The chapter mentions several characteristics of scientific knowledge. It also warns about confusing "commonsense" and "casual" observations with verified or potentially verifiable claims. With these considerations in mind, which of these statements would you consider to be an empirical claim? Which are normative statements? Which are so ambiguous that it's hard to tell? Write your responses in the space provided after each statement and briefly explain your answer.

a. The Republicans won a majority of seats in the Houses of Representatives in the 2010 midterm elections.

   Empirical – Count votes

b. Democrats are bad for the country.

   Normative – opinion

c. It is too bad for the country that the Republicans won a majority of seats in the House of Representatives in the 2010 midterm elections.

   Normative

d. Outlawing handguns will do no good because people will always find ways to kill each other.

   Ractopicet Ambigous

e. All people are created equal.

   Normative

f. Catholics are more likely to vote than are Protestants.

   Ambiguous *Empirical*

g. People in the Middle East would be far better off if they lived under democratic governments.

   Normative

h. Access to health care is a fundamental right.

   Normative

i. Abortion is always wrong.

   Normative

j. "The average 45-year-old high school dropout is in worse health than the average 65-year-old high school graduate."[1]

_____~~quantitative~~ empirical_____

_____

_____

_____

_____

k. In the United States it is less expensive to imprison a convicted murderer for life than it is to execute him or her.

_____~~empirical~~ empirical_____

_____

_____

_____

_____

l. It doesn't make any sense to vote because so many ballots are cast in an election that no single vote is going to make a difference in the outcome.

_____Normative_____

_____

_____

_____

_____

m. Because a study of marijuana showed no "substantial, systematic effect" on the brain, laws against this drug should be repealed.[2]

_____Normative_____

_____

_____

_____

_____

[1] Henry Levin and Nigel Holmes, "America's Learning Deficit," *New York Times*, November 7, 2005, 25.
[2] The study was cited in Brian Vastag, "Medical Marijuana Center Opens Doors," *JAMA* 290 (August 20, 2003): 878.

Exercise 2–3. Here are several paragraphs drawn from an article in which the author discusses a debate over how congressional districts are drawn:

I argue that map makers ought to "pack" districts with as many like-minded partisans as possible. Trying to draw "competitive districts" effectively cracks ideologically congruent votes into separate districts, which has the effect of increasing the absolute number of voters who will be unhappy with the outcome and dissatisfied with their representative.

One common objection to this method of districting [packing] is that it would add to the polarization in Congress by creating overwhelmingly Republican (Democratic) districts that are more likely to elect very conservative (liberal) members.

Some states, like Arizona, have passed laws or referenda specifying that a districting plan ought to maximize the number of competitive districts. This is not particularly surprising because the common wisdom among most voters and certainly among the media is that the House of Representatives does not have enough competitive districts currently, and that an increase in the number of competitive elections or in the amount of turnover in Congress will somehow enhance representation.

From: Thomas L. Brunell, "Rethinking Redistricting: How Drawing Uncompetitive Districts Eliminates Gerrymanders, Enhances Representation, and Improves Attitudes toward Congress," *PS: Political Science and Politics* 40 (January 2006): 77–85.

a. Identify two normative statements or claims from the preceding text that can't be tested empirically as currently expressed.

_____

_____

_____

_____

_____

b. Write down three statements or claims in the preceding text that are empirical and can be tested.

_____

_____

_____

_____

_____

Exercise 2–4. Many people make the following claim: "You can't predict human behavior." In light of our discussion of the scientific approach to political science, do you find this claim to be valid? (*Hint:* Try breaking "human behavior" down into more specific traits or properties. Example: "People are naturally aggressive." Then, think of ways that this might be empirically investigated.)

Humans are self Interested
- test children and tweens
to see if its true.

Exercise 2–5. "Traditional" political science was and continues to be criticized on several grounds. List them here.

Which of the criticisms do you find most compelling and why?

DECODING THE AMBIGUITY OF POLITICAL DISCOURSE

As we stated earlier, political discourse is frequently ambiguous, and you have to think carefully about what words really say. Sometimes a politician's meaning is clear, as when the president says, "Our country was attacked," which is a straightforward factual statement. But he also claims, "Our military was not receiving the resources it needed." Certainly in President George W. Bush's mind this, too, was a simple fact that could be demonstrated empirically. But the word *needed* makes the statement a judgment, not a factual proposition. Whether something is needed or not is an opinion. In some people's minds, the military has been receiving *more* than it needs, while others agree with the president that it has not. Who is right? It is hard to see how the proposition could be scientifically proven true or false.

Exercise 2–6. Visit http://www.whitehouse.gov/the-press-office/2011/07/29/remarks-president-fuel-efficiency-standards with your Web browser and read President Barack Obama's speech on fuel efficiency standards. Look for empirical generalizations, that is, statements that Obama intends to be taken as facts and not as his opinion. Which of those can you separate from normative assertions? Which statements purport to be factual or testable but are inherently nonempirical?

Exercise 2–7. Among the complaints lodged against "traditional" political science were that it overemphasized law and values, that it rested on subjective observation, and that it described particular events or institutions rather than explained general behavior. To get a sense of how the study of a particular topic has changed over time, read the following articles: Philip Marshall Brown, "The Theory of the Independence and Equality of States," *American Journal of International Law* 9 (April 1915): 305–35, and Bruce Bueno de Mesquita, "An Expected Utility Theory of International Conflict," *American Political Science Review* 74 (December 1980): 917–31. Do these articles follow the evolution of traditional into behavioral political science? In what ways? Most important, do modern studies help you understand the issues better than do the earlier ones? Which do you find most informative?

CHAPTER 3

Beginning the Research Process: Identifying a Research Topic, Developing Research Questions, and Reviewing the Literature

Probably everyone would agree that picking and narrowing a topic are the hardest tasks confronting a new researcher. One can, of course, easily identify problem areas such as the "war on terror" or "the effects of television on democracy." But moving from a desire to "do something on . . ." to a specific theme that can be researched with relatively few resources and little time can be quite challenging.

Part of the difficulty lies in having enough information about the subject matter. What is already known about it? How have previous investigators studied it? What important questions remain unanswered? All these considerations motivate the "review of the literature."

Chapter 3 of the textbook provides readers with some insights and tips into conducting an effective literature review. It is particularly important that you understand the differences between different kinds of sources, such as scholarly and mass circulation publications.

We assume that everyone knows roughly how to "surf" the Internet. So these assignments mainly force students to think carefully about what they are looking for and finding. As mentioned in the textbook chapter, you can easily enough use Google or equivalent software to search for "terrorism" or "television" or any other subject. But these efforts are usually unsuccessful because they lead to too much irrelevant information. Instead we encourage the application of more specialized databases and library tools.

Exercise 3–1. Find a copy of a newspaper that covers politics. On the lines following, write six research questions based on political news stories in the newspaper you selected. First write or attach the headline of the article, then the research question you created.

1. Jobs Invisible man strategy, will name ~~effect~~ *active voters*
2. Why is GOP going after security, does talking about social security affect votes?
3. Mitt Romney is scaring Hillary at facebook, does the fact that Hillary doesnt have a facebook hurt her chances
4. _____
5. _____
6. _____

Exercise 3–2.   Find a copy of a political science journal. Inside you will find a series of research articles. For five of the articles, identify the research question, explain where you located the research question in the article, and describe the political behavior the question focuses on (individual, group, and so on) and whether it is descriptive or asks about a relationship.

1. _____

2. _____

3. _____

4. _____

5. _____

Exercise 3–3.   Literature reviews are an important part of the research process. They provide the context and background so that a research project furthers our understanding of a political phenomenon by, among other things, attempting to resolve conflicting evidence, investigating a topic in different settings and populations, or using different measures of key concepts. Read the following excerpt of an article by David Niven.[1]

In reviewing the literature on the effects of negative campaign advertising, the author identifies several "problems" with the state of knowledge about the topic. What are these problems?

_____

_____

_____

_____

_____

_____

_____

_____

_____

_____

_____

_____

_____

_____

1. David Niven, "A Field Experiment on the Effects of Negative Campaign Mail on Voter Turnout in a Municipal Election," *Political Research Quarterly* 59, no. 2 (2006), 203–10. Excerpt from 203–5.

A Field Experiment on the Effects of Negative Campaign Mail on Voter Turnout in a Municipal Election

DAVID NIVEN, Ohio State University

This field experiment is used to expose a random sample of voters in a 2003 mayoral race to various pieces of negative direct mail advertising. Exposure to the negative advertising stimulus improved turnout overall about 6 percent over that of the control group. Results show that different topics and amounts of negative advertising had different effects on turnout. The results suggest that alarm bells sounded by some previous research and by public officials may be overheated, because the effects of campaign negativity may not be monolithic, and it would appear political negativity can have a positive effect on turnout.

Is voter turnout subject to the effects of negative advertising? Political science research answers alternatively yes, no, or maybe. This study uses a field experiment in which voters in a mayoral contest were randomly exposed to negative campaign mail to assess the effects of negativity and move toward a better understanding of what has become a thoroughly confusing line of scholarship.

Indeed two of the most prominent studies on campaign advertising offer quite differing views on the effects of negativity. Ansolabehere and Iyengar (1995) conclude that negative ads directly result in lower voter turnout. Far from qualifying their results, Ansolabehere and Iyengar (1995:12) assert the evidence is definitive that negative campaign messages "pose a serious threat to democracy" and are "the single biggest cause" of public disdain for politics (2). By contrast, Green and Gerber (2004: 59) describe the effect of campaign advertising negativity as "slight." Depending on the circumstances, Green and Gerber find negativity modestly nudging turnout upwards or downwards. Far from labeling their results conclusive, however, Green and Gerber suggest much more work needs to be done to better understand negativity's effect.

While this study addresses Green and Gerber's call for continuing research on this question, studying the effects of campaign negativity is of value beyond simply satisfying an academic curiosity. Understanding the effects of negativity obviously has implications for how candidates, parties, and interest groups conduct campaigns. Moreover, various government bodies have expressed interest in some form of negative ad regulation. Legislative proposals have been introduced at the local, state, and national level to limit negative campaigning with measures such as forcing candidates to appear in their ads or subjecting political advertising copy to some form of official scrutiny. Indeed, "I would ban negative ads," says Senator John McCain (R-Arizona) of the legislation he would create if he could find a constitutional procedure to accomplish the task.[1] Thus, to understand negativity and its effects better is to become better armed to participate in a debate which pits the First Amendment against the very popular notion of cleaning up campaigns.

NEGATIVITY AND ITS EFFECTS

While there is no consensus definition of negative advertising, most researchers start with the notion that negativity involves the invoking of an opponent by a candidate (for example, Djupe and Peterson 2002). That is, a negative ad suggests the opponent should not be elected rather than that the sponsoring candidate should be elected. West (2001) defines a negative campaign ad as advertising that focuses at least 50 percent of its attention on the opponent rather than the sponsor of the ad. Such negativity may be focused on any aspect of the opponent's record, statements, campaign, or background.

Precise estimates vary, but there is no doubt that negativity occupies a significant place in the modern campaign advertising arsenal. In the 2000 presidential election, for example, content analyses of television commercials from the two parties' nominees found between half and 70 percent were negative (Benoit et al. 2003; West 2001). Other forms of communication, such as radio ads, were even more negatively oriented (Benoit et al. 2003). Looked at from another tack, researchers have found as few as 20 percent of ads directed purely toward extolling the virtues of the sponsoring candidate (Freedman and Lawton 2004).

Employing a variety of methods, researchers have produced intriguing results in studies of negativity effects. However, those results variously demonstrate the negative, positive, or lack of effect of negative advertising on voter turnout.

Negative Ads Alienate Citizens

Dating back at least to the Watergate era, political scientists have documented the capacity of the American public to become categorically dismissive of political leaders. That is, the untrustworthy behavior of one political figure can transcend the individual and come to represent the political class as a whole (Arterton 1974; Craig 1993; Miller 1974).

---

[1] Quoted in Jennifer Holland, "McCain vows to keep campaign clean no matter what," Associated Press Wire Service, December 22, 1999.

Political Research Quarterly, Vol. 59, No. 2 (June 2006): pp. 203-210

Consistent with that notion, researchers have found evidence that negative political advertising negatively affects recipients' feelings not only toward the target of the attack but also toward its sponsor (Basil, Schooler, and Reeves 1991; Lemert, Wanta, and Lee 1999; Garramone 1984; Merritt 1984; Roese and Sande 1993) and even toward politics more generally (Ansolabehere and Iyengar 1995; Ansolabehere, Iyengar, Simon, and Valentino 1994; Houston and Roskos-Ewoldsen 1998; Houston, Doan, and Roskos-Ewoldsen 1999).

Using various real world races, including senate, gubernatorial, and mayoral campaigns, Ansolabehere and Iyengar (1995) exposed subjects in a laboratory setting to campaign television ads of various tone. Participants in Ansolabehere and Iyengar's experiments who were shown a negative television ad were almost 5 percent less likely to report they planned on voting in the upcoming election than participants who were shown a positive ad. Those who saw negative ads were also less likely to express confidence in the political system, and less likely to express political efficacy. Ansolabehere and Iyengar conclude that negativity in politics is causing declining voter interest and participation.

According to other experimental studies, the capacity for negative ads to produce diffuse political negativity varies with the precise details of the ads. For example, Budesheim, Houston, and DePaola (1996) found that unsubstantiated negative attacks reduced respondents' ratings of both the attacker and the target. See also Shapiro and Rieger (1992). Other scholars have suggested that issue related attacks are more apt to be seen as fair game than attacks focused on personal characteristics (Johnson-Cartee and Copeland 1989; Roddy and Garramone 1988).

Nevertheless, there is a significant limitation in experimental laboratory work on this subject that is inherent to the method. For example, Ansolabehere and colleagues show subjects' campaign ads then inquire about their *intention* to vote. Various other experimental studies inquire about intentions to vote, or candidate preferences, but none is equipped to measure actual resulting behavior. Of course, there is no shortage of psychological research demonstrating the gaping chasm between knowing someone's intentions or preferences and knowing their actual resulting behavior; for example, Kaiser and Gutscher (2003). Moreover, political scientists have regularly documented the propensity of Americans to mislead researchers when they are asked about their voting habits; for example, Bernstein, Chadha, and Montjoy (2001). Thus, regardless of the rigor of the researchers or the ingenious nature of their design, the laboratory remains a difficult setting in which to demonstrate the effect of negative advertising on the real world behavior of turning out to vote.

Negative Ads Do Not Alienate Citizens

Meanwhile, other researchers posit that the effects of negativity might not be negative at all. Finkel and Geer (1998), for example, argue that negative ads stimulate turnout because they provide highly relevant information.

Indeed, researchers have attributed positive or stimulating effects to feelings of negativity as an explanation for some notable political phenomena. For example, some scholars conclude that one source of the typical midterm loss, in which the president's party generally loses House seats in elections without the presidency on the ballot, is that voters who are critical of the president have a higher motivation to participate than voters who are positively inclined toward the president (Kernell 1977).[2]

Contemporary evidence also suggests that reception of negative advertising may contribute to effective citizenry. Brians and Wattenberg (1996), using survey data, show that citizens who recalled seeing negative political advertising during the 1992 presidential election were more accurate in assessing candidates' overall issue positions in that election.[3] In fact, recalling ads was more closely associated with holding accurate assessments of the candidates than was regularly watching television news or reading a newspaper. West (2001), studying the content of the ad rather than the effects on recipients, similarly supports the notion of the value of negative advertising. West (2001: 69) finds "the most substantive appeals actually came in negative spots."

Consistent with this line of thinking, several studies have found links between campaign negativity and increased voter turnout (Lau and Pomper 2001; Djupe and Peterson 2002; Kahn and Kenney 1999; Finkel and Geer 1998; Wattenberg and Brians 1999). Based on survey results or aggregate trends, these studies are better able than laboratory experiments to demonstrate actual voter turnout, but are far weaker in demonstrating individual reception of negative ads and thus are less firmly able to demonstrate a causal link between receiving ads and deciding to vote.[4]

Given the limitations of both laboratory experiments and non-experimental approaches, a strong argument can be made for the need for field experiments to address negativity effects. Field experiments offer internal validity (with random assignment and controlled exposure to the stimulus)

---

[2] A variety of psychological studies suggest the potential for superficially "negative" messages to have a "positive" effect on behavior. The implications of several lines of research considering the effects of showing people the negatives of such behaviors as cigarette smoking (Grandpre et. al. 2003) and motorcycle riding (Bellaby and Lawrenson 2001) find that simply demonstrating negatives is not an effective strategy in preventing participation. Indeed, the negative messages may draw attention and interest, and ultimately augment willingness to participate.

[3] Some have argued that the implied causality is backwards. That is, remembering ads does not encourage clear thoughts on issues, but having clear thoughts on issues does encourage remembering ads. See, for example, Ansolabehere, Iyengar, and Simon (1999).

[4] There is a further concern in aggregate studies. If candidates use negativity strategically, as we have every reason to believe they do (Theilmann and Wilhite 1998), then an accurate measure of campaign negativity may be, in effect, a proxy for some other variable affecting turnout. For example, Djupe and Peterson's (2002) data suggest that the amount of negativity in the U.S. Senate primaries they studied rose with the number of quality candidates. They attribute the resulting higher turnout to the campaign negativity, but surely an equally strong case could be made that the presence of more quality candidates was the true source of the turnout increase.

and external validity (with diverse participants and a measurement of the actual resulting behavior).

Relatively few field experiments on negative advertising have been reported. Pfau and Kenski (1990), did use field experiments to assess the strategic value of negative campaign messages by exposing randomly chosen voters to independently created direct mail and push poll messages. More recently, Green and Gerber (2004) have employed field experiments to study a vast array of potential campaign influences on voter turnout. Among their studies have been two which included negative political advertising sent by mail.

Green and Gerber (2004) sent negative campaign mail to a sample of voters in a Connecticut mayoral election. Here both reception of the ad and actual voter turnout can be established, and the subjects include a random sample of potential voters. Green and Gerber found the effects of negative ads on turnout in the mayoral race were negative but quite small. In another contest, using the same basic design but different mailings, they found the effect of negative ads on turnout was small but positive. Green and Gerber (2004: 59) tentatively conclude that the effect of negative campaign mail on turnout is best understood as "slight."

Why do Ansolabehere and Iyengar (1995) find negativity an inherent threat to voter turnout while Green and Gerber (2004) find negativity has little relevance to turnout? Differences in methodology could explain the disparate conclusions. Ansolabehere and Iyengar (1995) used television to convey negative messages while Green and Gerber (2004) used mail. However, nothing in Ansolabehere and Iyengar's (1995) theoretical approach suggests the effects of negativity require television as the medium of communica-

tion. Ansolabehere and Iyengar used a diverse but not random group of participants, while Green and Gerber (2004) used participants randomly drawn from several towns. However, nothing in Ansolabehere and Iyengar's (1995) protocol suggests they assembled a group of participants particularly attuned to the effects of negative messages. Probably the two most significant differences between the studies are that Ansolabehere and Iyengar's participants received their campaign communication in a laboratory, rather than in their homes (as was the case for Green and Gerber), and were asked about their intention to vote, rather than observed actually voting (as was the case for Green and Gerber). Both those factors might have contributed to an exaggeration of the negativity effect in Ansolabehere and Iyengar's study.[5] Beyond methodological differences, though, another compelling explanation exists. It is possible that both teams of researchers were measuring a realistic effect. That is, there may not be a monolithic negativity effect, and depending on the content of the ad and the circumstances of the race, negativity may in fact have quite varying effects on turnout.

Indeed, the confusing state of research in this area is well captured in Lau, Sigelman, Heldman, and Babbitt's (1999) meta-analysis of studies on negative ads. After building a weighty dossier of studies, both published and unpublished, they found that previous research findings suggesting negative ads increase turnout are available in similar quantity to findings suggesting negative ads decrease turnout. This leaves the authors to conclude that the cumulative estimated effect of all these studies of negativity on turnout approaches zero. It is, in short, an area which demands replication with the best methodological approach: a randomized field experiment.

[5] Ansolabehere and colleagues dispute the notion that their techniques exaggerated the effect of negativity. Indeed, they label their estimate of negativity's effect as "conservative" (Ansolabehere, Iyengar, Simon, and Valentino 1994: 835).

Exercise 3–4. Suppose you are working as a research assistant for a professor of political science who is beginning a new book about the current state of racial politics in the United States. She needs to make sure that she has read as much serious analytic writing as possible and wants you to begin compiling a bibliography of published materials. Which of these potential sources would you add to the list? Why? Which ones would be top priority and which ones would be a lower priority and why?

a. A book review by David Weakliem in the March 2004 issue of *Social Forces* of the book *The New Electoral Politics of Race* by Matthew Justin Streb

b. An article in *Reader's Digest* about Barack Obama

_____

_____

c. The book *Race, Politics, and Governance in the United States* (University Press of Florida, 1996) by Huey Perry

_____

_____

d. An article in the *Economist* on Barack Obama's quest for the Democratic presidential nomination

_____

_____

e. A book review by Franklin D. Gilliam Jr. in the winter 2000 issue of *Public Opinion Quarterly* of the book *Perception and Prejudice: Race and Politics in the United States* by Jon Hurwitz and Mark Peffley

_____

_____

f. An article in the August 1996 issue of *American Journal of Political Science* titled "A Racial/Ethnic Diversity Interpretation of Politics and Policy in the States of the U.S." by Rodney E. Hero and Caroline J. Tolbert

_____

_____

g. An editorial in the *Philadelphia Inquirer* titled "Immigration Isn't the Enemy; Inequity Is"

_____

_____

h. The book *Perception and Prejudice: Race and Politics in the United States* (Yale University Press, 1998) by Jon Hurwitz and Mark Peffley

_____

_____

i. The article "Black Electoral Power, White Resistance, and Legislative Behavior" by Gary H. Brooks and William Claggett, published in 1981 in the journal *Political Behavior*

_____

_____

j. An article by Ismail K. White published in the May 2007 issue of *American Political Science Review* titled "When Race Matters and When It Doesn't: Racial Group Differences in Response to Racial Cues"

_____

_____

k. An article in the *Washington Post* about Barack Obama's reelection campaign

_____

_____

Exercise 3–5. Suppose you want to write a term paper or scholarly report on one of the following subjects: "immigration policy in the United States," "nuclear nonproliferation treaty," "the World Trade Organization," or "climate change." Use one of these popular search engines—Google, Alltheweb, or Yahoo!—to begin building a bibliography.

a. Which search program did you choose?

_____

b. How many "hits" did your first search produce?

_____

c. How many sources on the *first* page of the search do you think would be helpful in writing an academic research report? Explain.

_____

_____

_____

_____

_____

d. Compile a brief bibliography—three citations for each of the following categories about your topic. See the textbook for a suggested format.

1. Articles in the mass media such as newspapers and magazines

_____

_____

_____

_____

_____

_____

2. Essays, reports, and discussions published on the Internet or elsewhere by advocacy groups, not-for-profit organizations, and government agencies

_____

_____

_____

_____

_____

_____

3. Scholarly articles

_____

_____

_____

_____

_____

_____

e. Now conduct a search using the other programs. Do these search engines generally locate the same sources, or are there important differences in what each finds? Which do you prefer? Why? And, more important, do you see the need to limit a topic?

_____

_____

_____

_____

_____

_____

_____

_____

_____

Exercise 3–6. Use Web of Science to find the following article: Nojin Kwak, Dhavan V. Shah, and R. Lance Holbert, "Connecting, Trusting, and Participating: The Direct and Interactive Effects of Social Associations," *Political Research Quarterly* 57, no. 4 (2004): 643–52.

a. How many sources did this article cite?

_____

b. What kind of sources were cited, and in what fields of study were the citations located?

_____

_____

_____

c. Click on each author's name. How many citations are listed for each? Do the topics of other work related to the article you are investigating appear? What evidence did you use to make this judgment?

_____

_____

_____

d. How many times has this article been cited?

_____

e. What kinds of work were cited this article, and in what fields of study were the citations located?

_____

_____

_____

f. Now click on the citation map to see a visual representation of how this article relied on previously published work and how subsequently published work relied on this article. How can a literature review help understand what is known about a topic of interest?

_____

_____

_____

Exercise 3–7.   Use Web of Science or Google Scholar to search for articles related to one of the following topics:

1.   Do harsh penalties have any effect on illegal drug use?
2.   Is the South more politically conservative than other regions?
3.   The role of civic culture in democracy
4.   Why do some states enact laws against gay marriage?
5.   Immigration to the United States from Latin America

Using an acceptable format, list your first five sources here.

1.   _____

2.   _____

3.   _____

4.   _____

5.   _____

Exercise 3–8.   Suppose you are interning in a law firm. One of the partners tells you that the firm has taken on a client accused of selling pornographic magazines and videos. Part of the accusation is that obscene video, printed, and Internet materials encourage male aggressiveness and dehumanize potential victims. The defense will claim that there is little or no empirical research to support this aspect of the charge. Obviously the partner needs to justify this position and asks you to compile a list of scholarly research on the topic. It is important, she tells you, to find experimental or quasi-experimental studies conducted by independent authorities. It does not matter what the investigators find; all that is important is that results appear in scholarly literature. More-over, her firm has already reviewed law journals and legal decisions. So you have to confine your investigation to social science (and possibly medical or health) sources. List ten of the reports you consider most relevant.

1.   _____

2.   _____

3.   _____

4.   _____

5.   _____

6.   _____

7.   _____

8.   _____

9.   _____

10. _____

Exercise 3–9. Sharpen your literature review skills by finding and listing six articles that rely primarily, or to some degree, on randomized experiments. A specific database such as JSTOR will be most rewarding. But here is a tip that can be applied to just about any kind of search: read *abstracts,* when available, instead of trying to skim entire articles. Using a correct format, list the results of your search here.

A TRICK FOR FINDING SOURCES

JSTOR may not be of much use because it does not yet archive many psychology journals, where a lot of the relevant literature is published. You could use Web of Science or some other database that includes articles from psychology journals. Or, instead, try this tactic: use the Internet to find papers that take a position one way or the other. Then, as described in chapter 3 of the textbook, consult the bibliographies and notes of the papers you find to "pyramid" further the list of sources. And, if in doubt about the references' appropriateness (for example, are they based on sound research conducted by reputable scholars and organizations?), try to track them down in the library.

1. _____

2. _____

3. _____

4. _____

5. _____

6. _____

Exercise 3–10. If you know of scholars who have conducted research on a topic of interest—perhaps you have culled names from the bibliographies in textbooks—you can search for their Web pages. Frequently, they will provide a curriculum vitae or list of publications, some of which may be in electronic form. Using the name of an author of one of the articles or book reviews mentioned in exercise 3–2, search for the author's Web page and print out his or her curriculum vitae or list of publications.

The Building Blocks of Social Scientific Research
Hypotheses, Concepts, and Variables

In chapter 4 of the textbook we explore some initial steps in the research process: how one may start with an interest in a political phenomenon or political concept, pose a research question about it, and propose an answer to the question in the form of a hypothesis. In this chapter we emphasize proposing suitable explanations for why a concept varies or how two concepts are related. A variable is a concept whose value is not constant; rather, it varies. Informed or suggested by theory or casual observation, hypotheses are "guesses" about relationships between variables. Hypotheses should be written so that the nature of the proposed relationship is clear, the concepts are distinct, and the unit of analysis is identified. Concepts or variables are attributes of an entity, something or someone, which is called a "unit of analysis." For example, units of analysis can be countries, cities, individuals, members of legislatures or courts, speeches, or government actions and activities.

Exercise 4–1.   Chapter 4 explains how the research question, theory, and hypotheses should all be consistent. The research question should pose an empirical question about the political world, the theory should provide a comprehensive answer to the research question, and the hypotheses should specify the relationship between two or more terms in the theory.

a.   In the space provided, write a research question, an outline of a theory, and at least three testable hypotheses.

_____

_____

_____

_____

_____

_____

_____

_____

_____

_____

_____

_____

_____

_____

_____

_____

_____

_____

_____

_____

b.   Specify the unit of analysis for your research question, theory, and hypotheses.

Exercise 4–2.   Following is a list of research questions. On the lines after each question, respond to these items:

■   Determine whether the research question seeks to investigate a relationship between concepts. If it does, go to the next item. If it does not, modify and rewrite the question as needed so that it proposes such a relationship, and then go to the next item.

■   Write a hypothesis that is consistent with the research question.

■   Identify the independent and dependent variables for the hypothesis.

■   Identify the unit of analysis indicated or implied by the research question and hypothesis.

IDENTIFYING VARIABLES AND UNITS OF ANALYSIS
An empirical hypothesis has three distinct components: the unit of analysis, the independent variable, and the dependent variable. Something cannot be both a unit of analysis and a variable. Something cannot be both an independent and a dependent variable.

D✓

a.   Is the US public more willing to support the use of US military forces when US interests are affected than when they are not?

Good

_____

Dv

b.   Are people around the world concerned about global climate change?

Are People In Industrial ~~Countries~~ concerned about climate change

_____

c. How important was the political experience of presidential candidates to voters in the 2008 presidential primaries?

[handwritten: DV over "important"; IV over "political experience"]

[handwritten answer:] Individuals,

d. Is the number of bills passed by a state legislature related to whether both chambers of the legislature are controlled by the same party?

[handwritten: DV over "number of bills passed"; IV under "controlled by the same party"]

e. Do states with public funding for candidates have more competitive elections than states without public funding?

[handwritten: IV over "public funding"; DV over "competitive"]

f. Does a country's turnout rate in elections have an impact on how much a country's welfare policies change in response to changes to market-based inequalities?

[handwritten: IV over "turnout rate"; DV over "welfare policies"]

Exercise 4-3. State how each of the following hypotheses could be improved and then rewrite the hypothesis.

a. There is a correlation between President Barack Obama's job approval ratings and public support for the war in Afghanistan.

[handwritten answer:] Direction, Be more specific

b. Higher murder rates are associated with higher school-dropout rates.

[handwritten answer:] Be specific about where

c. The turnout rate in my legislative district was low because the incumbent ran unopposed.

[handwritten answer:] make more generalizable

d. Cities with high rates of public transit ridership are better than cities in which ridership is low.

_____

_____

e. The more miles of coastline that a country has, the less likely it is to have a federal system of government.

_____

_____

f. People who are active in politics tend to contribute money to election campaigns more than people who are not active in politics.

_____

_____

Exercise 4–4. For each cluster of variables, write three hypotheses, pairing just two of the variables at a time. For each hypothesis, identify the independent variable, the dependent variable, and the unit of analysis.

a. Number of visits to political Web sites; political attitudes (liberal, moderate, conservative); years of education

_____

_____

_____

b. Child poverty rate within a state; births to unmarried women within a state; "toughness" of state welfare policy

_____

_____

_____

Exercise 4–5. Following are sets of three variables. On the lines after each set, respond to these items:

- Write a hypothesis relating the first two variables.
- Identify independent and dependent variables.
- State how you expect the third variable to affect the hypothesized relationship.
- Draw an arrow diagram including all three variables.
- Determine whether the third variable is antecedent, intervening, or alternative.

IV   DV

a. Primary caregiver for children (primary caregiver, not); support for Family Medical Leave Law (thermometer scale for support); gender (female, male).

IV

A female primary caregiver for children is more likely to support for family Medical leave law than male caregiver for childs. *Primary*

antecedent   intervening

female   primary   medical leave
           caregiver
           of kids

DV   IV

b. Intention to vote in upcoming election; respondent's general interest in politics; predicted outcome of election ("too close to call," "somewhat competitive," "lopsided victory")

Respondent's with a more general interest in politics intend to vote in upcoming elections

predicted outcome; antecedent
        ↓
interest   intervening
        ↓
Intention to vote

27

c. Type of light bulb purchased (regular, energy efficient); difference in cost of regular and high-efficiency lightbulbs (small difference, large difference); concern about global climate change

_____

_____

_____

CHAPTER 5

The Building Blocks of Social Scientific Research
Measurement

Measurement involves deciding how to measure the presence, absence, or number of concepts in a research project. Reliability and validity of measures are key concerns.

A reliable measure yields a consistent, stable result as long as the concept being measured remains unchanged. Measurement strategies that rely on memories, for example, may be quite unreliable, because the ability to remember specific information may vary, depending on when the measurement is made and whether distractions are present.

Valid measures correspond well with the meaning of the concept being measured. Researchers often develop rather elaborate schemes to measure complex concepts.

Level of measurement is an important aspect of a measurement scheme. There are four levels of measurement. From lowest to highest, these levels are as follows: nominal, ordinal, interval, and ratio. Choosing the appropriate statistics for the analysis of data depends on knowing the level of measurement of your variables. Frequently a variable can be measured using a variety of schemes. Choosing the scheme that uses the highest level possible provides the most information and is the most precise measure of a concept. Researchers frequently recode data, thus changing the level of measurement of a variable.

RECODING DATA
There are two strategies for recoding data to combine or collapse categories of a measure:

1. **Theoretical**. Choose categories that are meaningfully distinct, where theory would tell you that the differences between the categories are important or where you can see that there are distinct clusters of scores or values. For example, when combining actual household income amounts into income levels, a researcher might consider what the official poverty level is and group all households with incomes below that level into the lowest income group.
2. **Equally Sized Categories**. Choose categories so that each category has roughly an equal number of cases. In addition, limit the number of categories so that each category has at least ten cases.

Exercise 5–1. What is the level of measurement of the following measures? If you think that the level of measurement could be more than one type, explain your choice.

a. Employment sector (public, private)

    Nominal

b. State policy on gay marriage (banned by constitutional amendment, banned by legislative act, no ban)

    ~~Nominal~~   Ordinal

c. Marital status (never married, married, widowed, divorced, separated)

_____Nominal_____

_____

d. Child poverty (percentage of children living in poverty)

_____~~Interval~~   Ratio_____

_____

e. Number of hours spent on volunteer work per month (fewer than 5, between 5 and 10, more than 10)

_____Ordinal_____

_____

f. State income tax policy (no income tax, proportionate tax, graduated tax)

_____~~Nominal~~ Ordinal_____

_____

g. Voted in 2010 (voted, did not vote, ineligible)

_____Nominal_____

_____

h. Years of education (0, 1–8, 9–12, 13–16, 17+)

_____~~Ratio~~ Ordinal_____

_____

i. Political views (extremely liberal, liberal, slightly liberal, moderate, slightly conservative, conservative, extremely conservative)

_____~~Nominal~~ Ordinal_____

_____

j. Per-pupil education spending

_____~~Interval~~ Ratio_____

_____

k. Number of years served in Congress

_____~~Interval~~   Ratio_____

_____

l. Year first elected to public office

Interval ~~Ratio~~ *Interval*

m. How often one reads a newspaper (every day, 5–6 days per week, 3–4 days per week, 1–2 days per week, less than 1 day per week)

Ordinal

n. Tone of news article about a presidential candidate (positive, mixed, negative)

Nominal *Ordinal*

o. Year in college (freshman, sophomore, junior, senior)

Ordinal

p. Primary policy objective of military intervention (foreign policy restraint, humanitarian intervention, internal political change)

Nominal

Exercise 5–2. Suppose you are studying the extent to which various groups have experienced discrimination due to their racial and ethnic backgrounds. Which of the following questions (a or b) do you think would give you the most reliable responses? Why?

a. In your day-to-day life, how often do any of the following things happen to you because of your racial and ethnic background? Please give the number of times each of the following things has happened to you in the past three months.

1. You have been treated with less respect than other people.
2. You received poorer service than did other people at restaurants or stores.
3. People acted as if they were afraid of you.
4. You have been physically threatened or attacked.
5. You have been unfairly stopped by the police.

b. In your day-to-day life, how often do any of the following things happen to you because of your racial and ethnic background? Would you say very often, fairly often, once in a while, or never?

1. You have been treated with less respect than other people.
2. You received poorer service than did other people at restaurants or stores.
3. People acted as if they were afraid of you.
4. You have been physically threatened or attacked.
5. You have been unfairly stopped by the police.

_____

_____

_____

_____

Exercise 5–3. Herrmann, Tetlock, and Visser define the disposition of military assertiveness as "the inclination toward different methods of defending American interests abroad, in particular, whether a person prefers more militant and assertive strategies or more accommodative and cooperative approaches."[1] To measure military assertiveness, they used ten items. For the first eight items, they asked respondents to indicate whether they strongly agreed, agreed, neither agreed nor disagreed, disagreed, or strongly disagreed with the statement.

Which of the items following do you think are the most valid measures of the concept of military assertiveness and why? Which ones do you have trouble relating to the concept and why? What kind of validity (face or construct) do you think the items exhibit?

a. The best way to ensure world peace is through American military strength.
b. The use of military force only makes problems worse.
c. Rather than simply reacting to our enemies, it's better for us to strike first.
d. Generally, the more influence America has with other nations, the better off they are.
e. People can be divided into two distinct classes: the weak and the strong.
f. The facts on crime, sexual immorality, and the recent public disorders all show that we have to crack down harder on troublemakers if we are going to save our moral standards and preserve law and order.
g. Obedience and respect for authority are the most important virtues children should learn.
h. Although at times I may not agree with the government, my commitment to the United States always remains strong.
i. When you see the American flag flying, does it make you feel extremely good, somewhat good, or not very good?
j. How important is military defense spending to you personally: very important, important, or not at all important?

Most valid measures of the concept of military assertiveness

_____

_____

_____

_____

---

[1] Richard K. Herrmann, Philip E. Tetlock, and Penny S. Visser, "Mass Public Decisions to Go to War: A Cognitive-Interactionist Framework," *American Political Science Review* 93 (September 1999): 554.

Worst "fit" for concept

_____

_____

_____

Kind of validity

_____

_____

_____

_____

Exercise 5–4. Suppose that you think that moral values are theoretically important in explaining voting behavior. Before you can write your theory or test your hypotheses involving moral values, you must conceptualize and operationalize the concept. In the space following, conceptualize and operationalize moral values.

_____

_____

_____

_____

_____

_____

_____

_____

_____

Exercise 5–5. One strategy for measuring political ideology (strategy 1) is to ask respondents how important each goal is to them, using the five-category response scale of "very important," "somewhat important," "neither important nor unimportant," "somewhat unimportant," and "very unimportant." Answers could be scored and added together to create an index, with high scores indicating conservatism and low scores indicating liberalism. A second strategy (strategy 2) is to ask respondents to choose between goals as in the following items:

a. Narrowing the gap between the rich and the poor or helping the economy grow?
b. Encouraging belief in God or promoting a modern scientific outlook?
c. Working for the rights of women or preserving traditional family values?
d. Guaranteeing law and order in society or guaranteeing individual freedom?
e. Being tougher on criminals or protecting the rights of those accused of crime?
f. Defending the community's standards of right and wrong or protecting the rights of individuals to live by any moral standard they choose?

For each pair of choices, one choice would be considered the liberal choice, the other the conservative choice. The total number of conservative choices could be calculated to produce a liberal-conservative index score.

Which of these strategies, 1 or 2, would be better at distinguishing between conservatives and liberals? Why?

_____

_____

_____

_____

Exercise 5–6. Table 5–1 contains a frequency distribution of senators' scores on the American Federation of Labor–Congress of Industrial Organizations (AFL–CIO) rating system for 2005. The left-hand column shows the actual scores given to senators; the columns to the right show how many senators received the scores. Suppose you wanted to change these data into an ordinal-level measure with two categories. What are two ways this could be done? Give the range of scores that would fall into the categories.

_____

_____

_____

_____

_____

_____

TABLE 5–1

AFL-CIO 2005 Senators' Rating Scores

Score	Frequency	Percentage	Cumulative Percentage
0	1	1.0	1.0
7	16	16.0	17.0
8	1	1.0	18.0
14	19	19.0	37.0
15	2	2.0	39.0
17	1	1.0	40.0
21	6	6.0	46.0
23	1	1.0	47.0
29	3	3.0	50.0
46	1	1.0	51.0
50	1	1.0	52.0
57	3	3.0	55.0
64	1	1.0	56.0
69	1	1.0	57.0
71	3	3.0	60.0
77	2	2.0	62.0
79	10	10.0	72.0
85	1	1.0	73.0
86	7	7.0	80.0
92	3	3.0	83.0
93	8	8.0	91.0
100	9	9.0	100.0
Total	100	100.0	

Exercise 5–7. Table 5–2 shows the distribution of the average index scores of each state's delegation to the US House of Representatives on the League of Conservation Voters (LCV) index for 2006. The index ranges from 0 to 100 and represents the percentage of times that a member voted in favor of the LCV position on selected issues. Suppose that you wanted to group the average state delegation scores into four categories for a new variable called "Support for LCV." What range of values would be included in each of the categories? Justify your decision.

TABLE 5–2

League of Conservation Voters 2006 State Delegation Averages for the House

Score	Frequency	Percentage	Cumulative Percentage
0	3	6.0	6.0
4	1	2.0	8.0
5	1	2.0	10.0
9	1	2.0	12.0
14	1	2.0	14.0
19	1	2.0	16.0
20	2	4.0	20.0
22	2	4.0	24.0
25	1	2.0	26.0
26	1	2.0	28.0
27	2	4.0	32.0
32	1	2.0	34.0
36	1	2.0	36.0
37	1	2.0	38.0
38	2	4.0	42.0
39	4	8.0	50.0
40	1	2.0	52.0
42	1	2.0	54.0
45	1	2.0	56.0
47	1	2.0	58.0
50	2	4.0	62.0
52	1	2.0	64.0
54	1	2.0	66.0
61	1	2.0	68.0
62	1	2.0	70.0
67	3	6.0	76.0
68	1	2.0	78.0
75	1	2.0	80.0
77	1	2.0	82.0
78	2	4.0	86.0
83	1	2.0	88.0
85	1	2.0	90.0
88	2	4.0	94.0
89	1	2.0	96.0
99	1	2.0	98.0
100	1	2.0	100.0
Total	100	100.0	

Exercise 5–8.   Following you will find a series of hypotheses. For each hypothesis, identify the variables you would need to test the hypothesis and explain how you could measure each variable. When explaining your measurement strategy, be careful to consider validity and reliability.

a.   More religious people are more politically conservative than are less religious people.

b.   The farther voters have to travel to get to their polling station, the less likely they are to vote.

c.   As countries develop economically they are more likely to participate in peacekeeping missions.

d.   The frequency of major presidential addresses increases when the country is at war.

CHAPTER 6

Research Design: Making Causal Inferences

Chapter 6 of the textbook has two major goals. The first is to emphasize the importance of thinking through a research question in order to find methods and data that will throw light on the issue. The second is to describe an "ideal" standard of evidence against which results can be judged and to suggest some ways that researchers can strive to reach that level of verification.

On the surface, political scientists engage in all sorts of activities, few of which may look like causal analysis. But in that part of the discipline that thinks of itself as "scientific," a major goal—the Holy Grail, so to speak—is the search for verifiable causal relationships.

We began the chapter by showing what is necessary to demonstrate causality and how hard it is to do so. We also argued that the randomized experiment provides a model for supporting causal claims. Unfortunately, as powerful as they are, true experiments are neither feasible nor ethical in many research contexts. So we suggested alternative designs that might be called "approximations to experiments." By that phrase we mean procedures that accomplish roughly the same things as random assignment of subjects, physical manipulation and control of the test factor and experimental environment, and direct observation of measurement of behavior. But they do so indirectly and most often with statistics.

Most of the assignments following call for serious thought rather than paper-and-pencil calculations. The purpose is to ensure that some of the basic ideas are clearly understood. We do not expect that most students will be able to design and carry out a major empirical research project. At the same time, it is important to understand how systematic and rigorous research proceeds.

Exercise 6–1.   Identify whether each of the following statements makes a causal attribution or claim, merely states an association, or is too indeterminate to tell; briefly explain each response. Perhaps a few examples will help clarify what we are looking for. First, consider this statement: "Political radicals come from dysfunctional families." Here's a possible response. The statement may be totally false, but its meaning is relatively clear: a condition or factor, *X*, family background, causes *Y*, the development of extreme political attitudes and behaviors. Another example: "So-called 'hate crime' legislation would have a chilling effect on free speech. . . ."[1] This statement also makes a causal assertion, although it is very general and probably impossible to test in principle. (Why?) But the essence of the claim is that a condition, the passage of hate crime legislation, will create or bring about another condition, less freedom of speech. In the writer's mind, a causal "arrow" points directly from *X* to *Y*.

a.   Men are politically more conservative than women.

   Statement, Indeterminate

---

[1] Family Research Council, "Thought Crime ('Hate Crime') Laws: Unnecessary and a Threat to Free Speech." Available at http://www.frc.org/get.cfm?i=LH07D01&f=WX06K03.

b. The American-led invasion of Iraq and ouster of Saddam Hussein led to the chaos and bloodshed that engulfed the country after 2003.

Causal

c. Availability of birth control for teens increases the number of teens engaging in sex.

Causal

d. There will always be some people living in poverty no matter how hard government tries to eliminate it.

Statement, ~~Testable~~ Indeterminate

e. Voters who said health care was the most salient issue were more likely than other voters to have voted for Barack Obama in 2008.

Testable

f. "Marriage matters because it helps produce children who grow up to become responsible citizens. . . . Single-parent households are far more likely to produce children . . . who are dropouts, drug users, or criminals."[2]

Testable

g. "Show me a southerner, and I'll show you a political conservative."

~~Association~~ Statement

h. Indicate which of the preceding statements might involve a spurious or partially spurious correlation, and why.

_c._

Exercise 6–2. Chapter 1 brings up an important point about the judicial system in the United States. Recall that Jeffrey A. Segal and Albert D. Cover in one study and Jeff Yates and Andrew Whitford in another looked at how or on what grounds Supreme Court justices decide cases.[3] In particular, they asked whether justices rendered opinions solely on the basis of legal precedents and the application of law or whether other factors entered into the process. In the space following, draw three different diagrams that represent causal, spurious, and alternative propositions about judicial decision making. Also provide a short explanation of what the diagrams mean.

THINK ABOUT THE QUESTION
You do not need to know much about the Supreme Court to complete this exercise. First, reread the section in chapter 1 titled "A Look into Judicial Decision Making and Its Effects." Then, identify and list some of the variables or factors that are supposed to influence justices' opinions. (An obvious example is the degree to which a justice strictly adheres to legal principles. But the literature cited in the text mentions many others.) Finally, think about how these variables might be interconnected. Which of the connections would you call causal? Which are spurious?

---

[3] Jeffrey A. Segal and Albert D. Cover, "Ideological Values and the Votes of U.S. Supreme Court Justices," *American Political Science Review* 83 (June 1989): 557–65; and Jeff Yates and Andrew Whitford, "Presidential Power and the United States Supreme Court," *Political Research Quarterly* 51 (June 1998): 539–50.

Exercise 6–3. List the strengths and weaknesses of randomized controlled experiments.

a. Strengths

_____

_____

_____

_____

_____

_____

b. Weaknesses

_____

_____

_____

_____

_____

Exercise 6–4. An investigator wants to know if repeated and prolonged exposure to "pro-life" videos changes opinions about abortion policy. He draws a random sample of 100 people from the community of Nowhere and assigns them to one of four groups: The first 25 men to appear are assigned to group 1; the rest of the men are then placed in group 2; the first 25 women are assigned to group 3, and the remaining women go to group 4. (Groups 2 and 4 may have unequal numbers, but just ignore this possibility when answering.) The "treatments" are as follows:

Group 1 (Experimental): Over a period of three days the 25 male participants view 30 minutes per day of anti-abortion commercials.

Group 2 (Control): Over a period of three days the remaining males watch 30 minutes per day of automobile television advertisements and then go home.

Group 3 (Experimental): Over a period of three days the 25 females watch the same anti-abortion ads as the men for 30 minutes per day.

Group 4 (Control): Over a period of three days the remaining females see 30 minutes per day of automobile commercials and then go home.

The subjects' attitudes about abortion policy are measured at the time they are assigned to a group *and* one week after the last treatment has been administered. For the two experimental groups, the measurements show a large decrease in support for any kind of legal abortion. The control subjects did not change their views very much.

After collecting and analyzing all the data, the researcher arrives at two conclusions. First, exposure to persuasive messages does change opinions. In this case, the ads made both men and women less likely to support abortion rights. Second, the effect of the messages is exactly the same for men and women.

What do you think of this research design? In particular, answer the following questions:

a. Assume for a moment that the research design is sound. To what population, if any, can the results be generalized? (Write "yes" or "no" and a comment if you wish.)

    1. The people of Nowhere
    2. Only the men of Nowhere
    3. Only those people with heavy exposure to anti-abortion ads
    4. Men and women in the United States

b. What is external validity? In this context, which of these statements about external validity is correct?

    1. Men and women in the United States react similarly to persuasive communications in general.
    2. Men and women in the United States react similarly to the kinds of anti-abortion ads presented in the experiment.
    3. Men and women in Nowhere react similarly to these type of persuasive messages.
    4. Public opinion on political issues is affected by television communications.

c. Now go back to the design as presented. (Do not assume that it is faultless.) Is it sound? Which of the following research standards and principles seem to be violated? Explain.

Random sampling

_____
_____
_____
_____

Creation of a control group

_____
_____
_____
_____

Demand characteristics

_____
_____
_____
_____

Experimental mortality

_____
_____
_____
_____

Exercise 6–5.   This item is a question for discussion and study.

On March 23, 2010, President Barack Obama signed landmark health legislation into law. The health reforms were the signature legislative item on the president's domestic agenda. The bill was passed narrowly in both the House and Senate due to strong opposition from the Republican Party leadership in both chambers.

Suppose the White House asked you, as a research consultant, to answer the following questions. The president's staff wants to know if all the widely expressed doubts about health care reform are affecting the president's reelection chances in 2012. They are willing to fund your research but first need assurances that you can provide reliable and valid results. You therefore need to decide upon a research design or strategy that would best suit the problem.

What you want to know is how public support for the health law and the president changed over time after the bill was signed into law. In a brief response, answer the following questions: What type of research design would give you the best internal or external validity? How could you demonstrate that your results are reliable? What are the relative advantages of using a survey in a longitudinal or a panel design? Would a case study design be useful? Why or why not?

_____

_____

_____

_____

_____

_____

_____

_____

_____

_____

_____

_____

_____

_____

_____

Exercise 6–6.   Imagine that your professor mentioned in class that most college students rely on the *Daily Show* for political news and information. Your professor seems to think that following politics is necessary to get many of the jokes, but no one really learns anything by watching the show. You have decided to test your professor's ideas in the form of a research project for class.

In the space given, state the research question and hypotheses you intend to test as clearly and succinctly as possible. Next, explain how you would test these hypotheses and evaluate the strengths and weaknesses of this research strategy.

_____

_____

_____

_____

_____

_____

_____

_____

_____

_____

_____

_____

Exercise 6–7. For each of the following research questions describe the design or strategy you would propose to investigate the issue. Note that although you do not have to conduct the investigation, you should make your design as practical and moral as possible. Devise a strategy that you feel would maximize the "scientific" payoff.

a. What was the effect of the Gulf oil spill in April 2010 on President Obama's support among environmental interest groups?

_____

_____

_____

_____

_____

b. What were the social and political attitudes of the voters who supported the two major-party candidates in the 2008 presidential election?

_____

_____

_____

_____

c. What are the effects of postregistration policies (for example, mailing to each registered voter a sample ballot and information about polling places and times, extended poll hours, and making election day an official holiday) on voter turnout in elections?

_____

_____

_____

_____

_____

d. Why do some states have the death penalty whereas others do not?

_____

_____

_____

_____

_____

e. What is the effect of ballot position on voter choice in a city election in which there are no incumbents and no party labels?

_____

_____

_____

_____

_____

f. What might be the effect of a 50¢-a-gallon increase in the price of gasoline on the use of public transportation?

_____

_____

_____

_____

_____

g. What was the reaction of the leadership of the Mexican political party PRI (Partido de la Revolución Democrática) to the passage of the North American Free Trade Agreement?

_____

_____

_____

_____

_____

h. Can imposing a refundable deposit (such as applied to glass and plastic bottles in some states) on each new tire sold in a state help reduce the number of used tires dumped illegally?

_____

_____

_____

_____

_____

Exercise 6–8. Think about the claim "Marijuana is a 'gateway' drug." Presumably the term *gateway* means that once a person starts using marijuana products, there will be some probability that he or she will try other drugs and that this probability will be greater than for a person who does not use such products. Marijuana is, in other words, a probabilistic cause of future drug abuse. Suppose the Drug Enforcement Agency (DEA) has asked your consulting firm to test this claim. First, briefly outline a randomized experiment that might throw light on the question. Then, discuss the possible ethical issues that could arise in this research. (Since you are under contract with the DEA, don't worry about the legality of obtaining the drugs.) If there are ethical problems, can you think of alternatives to experimentation that would allow you to make causal inferences?

_____

_____

_____

_____

_____

_____

_____

_____

_____

_____

_____

_____

_____

Exercise 6–9.   For this assignment you will first need to read Richard Fenno's 1977 article, "U.S. House Members in Their Constituencies: An Exploration," *American Political Science Review* 71: 883–917. You will want to pay special attention to the first ten to twelve pages. As you will see, Fenno relied on a participant observation (see chapter 8) research design in this article. For this exercise consider and describe how you could test Fenno's conclusions with an alternative research design.

a.   What type of research design would you use?

_____

_____

_____

b.   What are the advantages and disadvantages of this research design relative to Fenno's design?

_____

_____

_____

c.   How would you operationalize a homestyle?

_____

_____

_____

CHAPTER 7

Sampling

The attempt to verify statements empirically lies at the core of modern political science. Abstract theorizing is a valuable, even necessary, activity. Still, most social scientists feel that at some point theories have to "face reality." Thus, as we have seen in several chapters (like chapters 4, 5, and 6), carefully observing and collecting data is an integral part of the research process.

Unfortunately, in all too many situations it is not possible to observe each member of a population. Hence, sampling—the process of drawing a small set of cases from a larger population—becomes necessary.

The social sciences depend heavily on sampling. This fact sometimes troubles the general public. "How," many citizens ask, "can you make a claim about all the 290 million people in the United States when you've interviewed just five hundred of them?" Still, some people, including many reporters, politicians, and political advisers, act as though polling is an exact science. Chapter 7 addresses these issues.

More specifically, sampling raises two questions. First, *how* should the subset of observations be collected from the population, and second, *how reliable and valid* are inferences made on the basis of a sample? The first question pertains to sample types or designs, whereas the second deals with statistics and probability.

At this level of your training it is not possible to go into detail about either question. But if you work through these assignments you may begin to get a feel for the basics of sampling techniques and their properties. None of them involves any mathematical sophistication. But they do require careful thought.

Exercise 7–1. The following is an excerpt from an article by Laura Beth Nielsen titled "Situating Legal Consciousness: Experiences and Attitudes of Ordinary Citizens about Law and Street Harassment."[1] Read the excerpt and answer the questions that follow.

---

[1] Laura Beth Nielsen, "Situating Legal Consciousness: Experiences and Attitudes of Ordinary Citizens about Law and Street Harassment," *Law and Society Review* 34, no. 4 (2000): 1055–90.

IV. Method

The empirical study of legal consciousness presents several methodological challenges. Legal consciousness is complex and difficult to inquire about without inventing it for the subjects, or, at the very least, biasing the subjects' responses. Only through in-depth interviews can legal consciousness emerge, leaving the researcher with lengthy transcripts and the daunting task of using them to determine how to gauge variation in legal consciousness and how this relates to broader social structures.

Early studies attempted to capture the complexities of legal consciousness through observation and in-depth interviews with small samples (see Ewick and Silbey 1992; White 1991; Merry 1990; Sarat 1990). These methods were necessary as theories of legal consciousness were developing. More recently, scholars of legal consciousness have begun to advocate broader data collection to understand variation in legal consciousness and to map the relationship between consciousness and social structure (McCann 1999; Ewick & Silbey 1998; McCann & March 1996). In contrast, studies of political tolerance have surveyed large, randomly selected samples of citizens. The structured protocols of this line of research document attitudes and opinions, but do not allow for an in-depth understanding of legal consciousness.

I bridge this gap by using qualitative research techniques to probe the complexity of legal consciousness, while also interviewing a large enough number of subjects of different races, genders, and classes ($n = 100$) to begin to gauge variation in it. The combination of field observation and in-depth interviews proved especially valuable. The field observations allowed me to witness and record various types of interactions between strangers in public places. Because I observed many subjects being harassed in public and their reactions to such comments, I was able to guard against the tendency some subjects might have had to inflate the bravado with which they responded to such comments. Of course, simply observing was not sufficient because I needed to learn how the subjects *experienced* such interactions, not simply how they responded. The in-depth interviews provided an opportunity to gain an understanding of how individuals think about such interactions, resulting in a "mutuality" between participant observation and in-depth interviews (Lofland & Lofland 1995).

I systematically sampled subjects from the public places I observed. This strategy has several advantages. First, I knew that the subjects were consumers of public space, and thus they constituted a set of potential targets for offensive public speech. From my observations at different locations at different times, I also had some appreciation for what the subjects experienced. Second, by approaching subjects in person, I could establish rapport in a way that would have been impossible if I had initiated contact by telephone. This rapport was essential, given the sensitive nature of the interview questions. Asking subjects about experiences with offensive racist and sexist speech required speaking bluntly and using racial epithets as examples. It would have been difficult to gain consent without such personal contact.

I followed systematic procedures to construct a sample that, while not a probability sample, included different types of people and minimized the possibility of researcher-biased selections because of my personal prepossessions and characteristics. Of course, my presence in the public spaces might have altered the nature of the interactions that took place. Yet in most instances I was simply another person in the crowd and did not have much impact on the obvious interactions taking place.

I conducted a detailed assessment of data sites with the objective of maximizing variation in the socioeconomic status of potential subjects and guarding against idiosyncratic factors that might bias the results (Lofland & Lofland 1995). First, I selected field sites in a variety of locations in three communities in the San Francisco, California, Bay Area (Orinda, Berkeley/Oakland, and San Francisco) to insure broad representation across race, socioeconomic status, and gender among subjects selected to participate in the interviews. Second, I varied the day of the week, going to each of the locations on weekdays and weekends. Third, I varied the time of day by observing in each location during day, evening, and night hours. The field sites I selected were public places, such as sidewalks, public transportation terminals, and bus stops. Finally, to guard against approaching only potential subjects with whom I felt comfortable and to randomize subject selection within field sites, I devised a system whereby each person in the site had an equal chance of being approached.[8] I selected individuals to approach and asked whether they would participate in an interview about interactions among strangers in public places. I continued such selections until I achieved numerical goals for respondents with certain racial and gender characteristics. I oversampled white women and people of color for analytic purposes. Thus, even though I randomized selections within demographic subgroups and within strategically selected locations, this was not a random sample.[9]

---

[8] When I entered a field site, I recorded the scene in my field notes, noting the date, time of day, location, and characteristics of the people occupying that location. I also noted all instances of street harassment. I observed interactions, noting the types of individuals who made comments to strangers, and what responses they received. To determine whom to approach, I randomly selected a side of the location (north or south, east or west) by the flip of a coin. I rolled a die to determine the interval among individuals I would approach; for example, if the coin came up "heads" I went to the north side of the location (such as a train platform); then, if the die came up "3," I approached every third person to ask if he or she would be willing to participate.

[9] In the analyses that follow, I emphasize comparisons across race and gender and limit the statistical analysis to simple chi-square tests for differences across groups. Given the size of the sample, the results should be seen as suggestive in a statistical sense and worthy of examination in larger sample designs.

a. What type of sample did the author use?

_____

_____

b. How did she try to limit bias in her sample?

_____

_____

_____

_____

_____

_____

c. How did the author's method of selecting her subjects ensure that they were appropriate for her study?

_____

_____

_____

_____

_____

_____

Exercise 7–2. Consider this hypothesis: High school students have political beliefs and attitudes similar to those of their parents. Both students and parents will be sent questionnaires and their responses compared. The work will be done at "South High," which has an enrollment of 2,000. Here are some ideas for collecting the data. In each instance identify the sampling design and indicate whether it would produce data for a satisfactory test of the hypothesis. Briefly explain. To what populations, if any, could the results be generalized? (*Note:* Do not worry about aspects of the project such as how questionnaires will be matched or obtaining permissions from the school board and others.)

a. *Proposed sampling scheme:* The investigator takes the first and last names (and addresses) from every other page of South High's student directory and mails a questionnaire to those students and their parents.

Systematic sample, Generalized, Satisfactory
Structure, Parents South high

b. *Proposed sampling scheme:* Beginning March 1 at 3:30, the investigator stands outside the entrance to South High and hands out questionnaires to passing students and asks that they and their parents return them.

Convenience Sample - non Probability
Not good

c. *Proposed sampling scheme:* Investigator asks South High's assistant principal to generate a random list of 200 student names and addresses. Each of these students and his or her parents are mailed a questionnaire.

Simple Random Sample

d. *Proposed sampling scheme:* The investigator asks the guidance counselor for the names of exactly 50 college-prep students, 50 general study students, 50 vocational educational students, and 50 other students of any kind.

Stratified sample   doest
Quota, not random
have to be

e. *Proposed sampling scheme:* The investigator asks South High's assistant principal to draw (randomly) 50 names from each class (freshman, sophomore, junior, and senior). Each of these students and his or her parents are mailed a questionnaire.

Stratified Sample – Has to be random

Exercise 7–3. Suppose you work in the governor's office in Maryland. You have been asked to compare the experiences of businesses owned by various ethnic groups with respect to their interaction with the state economic development office. Because all businesses must register with your state, you have a current list of all businesses and their addresses. Unfortunately, your information does not contain data about the ethnicity of the owners. You plan to send a questionnaire to a sample of business owners. Now the question of sample size comes up. Your office has limited funds but needs to make reliable inferences. Fortunately, US Bureau of the Census data indicate the percentage of firms owned by various ethnic groups as shown in table 7–1.

TABLE 7-1

	Maryland Firms, by Race, 2002		Expected Numbers for Samples of	
Group	Population	Percentage	200	1,000
White	329,107	74.2		
Black	69,192	15.6		
American Indian/ Alaska Native	3,548	0.8		
Asian	26,169	5.9		
Hispanic	15,524	3.5		
Total	443,540	100.0		

Source: Adapted from US Bureau of the Census, *State and County Quickfacts.* Accessed November 1, 2007, from http://quickfacts.census.gov/qfd/states/24000.html.

a. If you conduct a total simple random sample of 200, what is the expected number of businesses in each ethnic group? Round to the nearest whole number. Write them in the table.

b. What about a sample of 1,000? Enter these expectations in the last column.

c. Do you see any problems with the sample sizes? Explain.

a. Now assume that business registration forms contained information about the ethnicity of the owners. How would you take a probability sample of 200 owners?

_____

_____

_____

_____

_____

Exercise 7–4. To continue the previous case, suppose that to save money you wanted a total sample of 200 Maryland business owners. But to make sure the ethnic groups were represented by adequate numbers, you conducted a disproportionate sample and found the mean (average) number of employees in each group. The sample sizes and average number of employees per group are shown in table 7–2.

TABLE 7-2

Group	Final Sample Size	Mean Number of Employees	Sampling Fraction	Weighting Factor	Weighted Mean
White	70	322			
Black	40	65			
American Indian/ Alaska Native	30	15			
Asian	30	27			
Hispanic	30	30			
Total	200				

a. What is the sampling fraction for each group? Write answers in the table. (*Hint:* Ask what proportion of a group leads to a sample of 70, 40, or 30, as the case may be.)
b. Suppose you found the sample mean (average) number of employees for each ethnic group (see table 7–2). For each ethnic group, calculate its weighting factor that should enter the calculation of an overall estimate of the mean number of employees. Put those numbers in the last columns.
c. What is the *overall* weighted mean? (*Hint:* If the sample is disproportionate, members of some groups are by definition over-represented. So they should "count" for less when calculating the average. How did we do it in the text?)

_____

_____

_____

_____

_____

_____

_____

Exercise 7–5. Following are some organizations and companies that do extensive polling. It is instructive to see how they conduct this research. It is especially interesting to compare not-for-profit or academic surveys with commercial polls. For each organization or company, locate and identify the sample type and size it uses in polling. Are there any differences between commercial and noncommercial surveys with regard to the types and sizes of samples used? (*Note:* By the time you read this some of the Internet addresses may have changed. But for the most part these are well-established programs, and by searching for their names you should be able to locate the needed information quickly.)

a. Chicago Council on Global Affairs
    American Public Opinion and Foreign Policy
    http://www.thechicagocouncil.org/ (click on "Studies and Publications," then "Foreign Policy Studies and Public Opinion")

    Typical sample type (e.g., telephone, face to face): _____

    Typical sample size: _____

b. National Opinion Research Center (NORC)
    General Social Surveys
    http://www.norc.org/projects/General+Social+Survey.htm

    Typical sample type (e.g., telephone, face to face): _____

    Typical sample size: _____

c. National Election Studies (NES)
    http://www.electionstudies.org

    Typical sample type (e.g., telephone, face to face): _____

    Typical sample size: _____

d. The Pew Research Center
    http://people-press.org

    Typical sample type (e.g., telephone, face to face): _____

    Typical sample size: _____

e. *Washington Post* Online
    http://www.washingtonpost.com/wp-srv/politics/polls/vault/vault.htm

    Typical sample type (e.g., telephone, face to face): _____

    Typical sample size: _____

f. CBS News/*New York Times* Polls
    http://topics.nytimes.com/top/reference/timestopics/subjects/n/newyorktimes-poll-watch/index .html?8qa

Typical sample type (e.g., telephone, face to face): _____

Typical sample size: _____

g. RealClearPolitics
(Offers links to many polls. Select a polling organization not listed previously.)
http://www.realclearpolitics.com/polls

Typical sample type (e.g., telephone, face to face): _____

Typical sample size: _____

h. As a rule, do you see any noticeable differences in sampling methods and sizes?

_____

_____

_____

_____

_____

_____

Exercise 7–6. Assume that .158 is the true proportion, P, of the people who do not have health insurance in the United States. If you took a sample of 400 people living in the United States at that time:

a. What would be the expected value of your estimator for this proportion?

_____

b. Assume standard assumptions hold. What would be the standard error of this estimator?

_____

c. Suppose the sample size was increased to 1,000. What is the expected value of your estimator?

_____

d. What is the new standard error?

_____

e. Suppose a survey of 1,000 citizens finds that the proportion of respondents who report not having health insurance is .200. Would this sample result be unusual if the true proportion in the population is .158? Explain. (*Hint:* Try to draw a crude bar graph [such as that shown in figure 7–4 of the textbook], locate the *population* value under the peak bar, and then think about what the sample result suggests.)

_____

_____

_____

_____

Exercise 7–7. The purpose of this assignment is to illustrate sampling procedures and statistical estimation.

Assume that you have been asked by a nonprofit advocacy group with limited funds to conduct a simple poll. It wants to know the proportion of people in Dullsville (a town of 5,000 adults) who would oppose the establishment of a halfway house for women who have recently been released from prison. The group believes that if 50 percent or fewer oppose the idea, it can begin negotiations with city officials on creation of the halfway house. If, however, more than 50 percent are against the facility, the group may look for a site in another community.

Your task is to draw a random sample of ten people from the population of residents and estimate the proportion that is against locating the facility in the community.

For this exercise you need to download two files from the Web site (http://psrm.cqpress.com): one contains a table of 10,000 random numbers; the other file is an enumeration of the "population." The 5,000 adults living in this community have been numbered consecutively from 1 to 5,000, and each person's response to a survey asking about the halfway house has been recorded as "For" (0) or "Against" (1) the proposal. In a nutshell, you will pick ten numbers from the random number table and then locate those specific individuals in the survey file and record their responses. The proportion (or percentage) of those in your sample who are against the house is your estimate of the population proportion.

By the way, your instructor knows the true value, *P.* This knowledge presents you with an interesting challenge, one that goes to the heart of statistical inference. If, like some students, you fall behind in your work, you might be tempted to "wing" this assignment by just making up and reporting a percentage without actually drawing a sample. The problem is that your guess has to be pretty good; otherwise, the instructor might infer that your estimate is so far off the mark that it couldn't have come from a truly random sample and that you must have cut corners. Suppose, for instance, your estimate of *P* is .1, whereas the true value is .8. This discrepancy might be grounds for doubting whether you did the work honestly. Still, and here is your problem, there is a chance, albeit very slight, that your sample result really does turn out to be .1, when the true value is .8. If you were "convicted" in this rare situation, the instructor would be making a mistake. Your knowledge of statistics may help you defend yourself.

UNDERSTANDING DATA FILES

Neither the textbook nor this workbook offers much instruction in using computer software to analyze data. Many software packages are available, and political scientists have not adopted a standard. Your instructor will guide your use of the program adopted for the course.

Nevertheless, most software works the same way, and we can provide a few general tips that may be helpful for getting data into a program such as SPSS.

File Extensions. Information is stored electronically in different formats. You can often tell the format by looking at the file name and especially at the *file extension,* which comprises a period and three letters. Knowing the file format lets you pick the correct program or program options when reading or opening a data file. Some common file extensions are as follows:

- **.txt** for "text" data or information. A text file contains just alphanumeric characters (for example, letters, digits, punctuation marks, a few symbols) and, when printed, looks just like something created on a typewriter or simple printer. If a program "thinks" it's reading text data, it won't recognize hidden codes for different fonts, graphics, and so forth. Consequently, if your word processor or editor (for example, NotePad) shows you a lot of gibberish, chances are that the file is not simple text. When you double-click a file name of this sort, your operating system's default word processor or editor will automatically try to open it. An example of a text file is "anes-2004readme.txt," which describes a set of data pertaining to the 2004 American national election.

- **.dat** for "data." The extension does indeed suggest data, but files of this type sometimes contain alphanumeric characters as well. In either case they can be loaded into a word processor. Moreover, some statistical programs recognize the ".dat" extension as data and will try to open the data. SPSS, for example, reads these files. Go to "File," and then "Read text data." After locating the file in the menu box, the program will start a "Text Import Wizard," which takes you step by step through getting the data. Examples of this format are "randomnumbers.dat," "surveytext.dat," and "surveydigits.dat." Depending on your system's configuration, double-clicking on ".dat" extension names will start a word processor or possibly a statistical program. But you can first run the program you want and then read the file.

- **.doc** for "document" information. This extension usually means Microsoft Word–formatted information that contains hidden formatting codes and so forth. Unless you have changed options on your computer or do not have the Windows operating system, double-clicking a ".doc" file will start Microsoft Word. (As we mention in the text following, other word processors can open some versions of Word files, so you are not limited to just that package.)

- **.sav** and **.por** for SPSS data files. These file extensions "belong" to SPSS. Like most statistical software programs, SPSS allows you to give descriptive names to variables and their individual values and to create new variables or transform and recode variables in all sorts of ways. All this auxiliary information along with the raw data can then be saved in one file so that it is available for reuse at a later time. The file extension ".sav" stands for "saved." SPSS data; dictionary information can be saved in a slightly more general format called ".por" for "portable." (We frequently use this option.) These files can be read by SPSS running on operating systems like Unix. An example is "surveydigits.por."

File Structure and Size. The file structure we use is quite simple: data are presented and stored in rectangular arrays in which each row represents a case (an individual, for instance), and the columns contain values of the variables. So if a file has 1,000 cases and two variables, the data structure is a 1,000-by-2 rectangular array of cells. Each cell holds a value for a specific case for a specific variable. (*Note:* To save space on the printed page, we sometimes use "unstacked" columns.) That is, the "surveydigits.dat" file, for example, has five columns of identification numbers and five columns of responses to make a 1,000-by-10-column matrix. But we arranged the numbers this way purely for convenience. Most software lets you stack columns on top of one another. Therefore, if you wanted, you could stack the columns 1, 3, 5, 7, and 9 of "surveydigits.dat" on top of each other and do the same with columns 2, 4, 6, 8, 10, to make a 5,000-by-2 array. Notice in addition that files with more than, say, thirty variables need more than one line when shown on a monitor or printed on an average piece of paper. For these data sets the lines will "wrap" around, making them difficult to read. Finally, if you are thinking about copying a file, you can roughly estimate the file's size by multiplying the number of variables by the number of cases.

File Delimiters. Most of the time the data points are separated by simple blank spaces. Occasionally, however, data are separated by "tabs." (In many systems the tab character is denoted by "^t," that is, a caret and lowercase *t*.) Sometimes you have to keep this in mind when using certain software, but many times a program will detect the tabs automatically.

Case ID Numbers. Some data files have explicit identification numbers for each case (for example, "surveytext.dat" and "survey digits.dat"). In others the case number is just the row number. When you view the data matrix in a program, you will be able to determine which is the case.

Note: Our Web site (http://psrm.cqpress.com) contains all the data files.

In any event, to get you started, here are some more specific instructions. First, open the table of random numbers and copy it to your word processing program. (You could print it out but there's no need to.) There are two formats for this file, a plain text version, "randomnumbers.dat," which can be loaded into any program, such as Word, WordPad, or WordPerfect, and a second version, "randomnumbers.doc," stored in Microsoft Word format.[2] Whichever one you decide to use, pick an arbitrary starting place in the table and copy ten consecutive numbers in the spaces of the blank table following. (*Note:* If you find a number larger than 5,000 (for example, 6,889), just skip it and go to the next.) These are your sample identification numbers. Next, open the survey data file. It too comes in a couple of formats: "surveytext.dat" and "surveydigits.dat" are text files that can be read by most word processors. In the first the responses are recorded literally as "For" and "Against." The second codes the responses as 0 for "For" and 1 for "Against." (There is also a "portable" SPSS file, "survey-digits.por," which contains the numeric codes.)

Naturally, these files create an unrealistic situation. If you had the entire population's responses, you wouldn't need a sample. But let's pretend that the total population is inaccessible, as in real life, and that you can only draw a sample from it. So using either the text or digits file, search for the first respondent ID number and write his or her response in the blank table following. Then, locate the second person, record the answer, and continue until you have all ten sample values. (*Hint:* Let your software program's "Find" function help you locate the respondents. In most systems, press "Ctrl-F." Thus, if a number is 2,493, press "Ctrl-F," enter "2,493," and press "OK." You should go right to that case. Note that if you are looking for individual 213, make sure you find "213" and not, say, "2,130.")

a. Write the randomly chosen identification numbers and responses in the blank table.

Respondent Number	Response

b. What is your estimate of the proportion *against* the halfway house?

_____

_____

---

[2] Several word processors will open files stored in a competitor's format. Hence, this version should be usable in many different situations. For example, recent versions of WordPerfect will read Word .doc files.

c. Write a one- or two-paragraph report for the group describing your results. Do you recommend going forward to the city council? Why? (*Note:* Reread the "Sampling Distributions" section in chapter 7.)

_____
_____
_____
_____
_____
_____
_____
_____
_____
_____

Exercise 7–8. This assignment continues exercise 7–7, but this time you will draw a *systematic* random sample of ten identification numbers.

a. For this situation you have to select every   observation. (This is the sampling interval).
b. Choose a random starting point. What number did you get? (*Note:* If your start number is greater than 5,000, go to the next number or select another random starting place.)

_____

c. Select the first case and record its response in the following blank table.

d. Draw nine more cases by incrementing the previously selected number by the sample interval. Record the responses here. (After getting a starting number, just keep adding the increment calculated in step 7–8a to get the next number. Continue in this fashion until you are done. You don't literally have to count down the pages line by line until you get to the next number.)

Respondent Number	Response

e. For this sample, what is the estimated proportion against the proposed house?

_____

f. Since both samples are random, combine the results to obtain an overall estimate of P and make a recommendation on that averaged value. Should the organization go ahead with its plans?

_____

_____

_____

_____

_____

_____

_____

_____

_____

_____

Exercise 7–9. Let's try to obtain an intuitive feel for sampling, estimation, and sampling distributions by drawing ten more independent random samples from the population represented in the survey files and calculating a proportion for each. Load the file "surveydigits.dat" into a statistical program that can draw random samples from rows. If you have access to SPSS, use the "surveydigits.por" file, open the data menu, and click "Select Cases, Random sample of cases" to obtain samples from the columns. You will have to repeat this procedure ten times to get ten samples. Other programs allow you to copy a randomly selected set of row values into new columns. Or, with a little bit of effort, you can obtain the samples manually as you did in the previous exercises.

a. List the responses and estimated proportion of people against the halfway house in the blank table.

Sample Number									
1	2	3	4	5	6	7	8	9	10

b. What is the mean or average of these ten sample proportions?

_____

_____

c. What statistical principle does this average illustrate?

_____

_____

Exercise 7–10. You are working for a nonprofit agency that is trying to encourage political mobilization in Africa. At the moment the organization plans to conduct a study in a largely rural nation on citizens' attitudes toward participation and elections. There are no lists of voters. But a population census has recently been conducted, and the numbers of people living in each province, district, and subdistrict are known to a reasonable degree of accuracy. Moreover, maps of these areas show detail down to block-sized segments. Given this information, propose a sampling strategy. Since the data are intended for statistical inference, you want the selection of people to be random. But of course a simple random sample is not possible. So what would you suggest as an alternative?

_____

_____

_____

_____

_____

_____

_____

_____

_____

_____

Exercise 7–11. In this exercise, you will be simulating sampling distributions. (Seeing is believing.) Chapter 7 on sampling and later chapters covering statistics (chapters 11–14) make frequent reference to sampling distributions. The basic idea is fairly simple but a bit abstract. So this exercise allows you to see what is meant by a sampling distribution being the result of a very large number of independent samples from a population. The simulation is located at "Sampling Distributions," Rice Virtual Lab in Statistics (http://onlinestatbook.com/statsim/sampling_dist/index.html).

In brief, this exercise asks you to draw samples from a known, normally distributed population with a mean (μ) of 16 and a standard deviation (σ) of 5. The size of your samples varies. Initially, you draw just two cases and find their average. Then you sample another two observations and determine their mean. This process continues until you have a batch of means, each of which is based on just two observations. When you are done you will see that the average of these sample means is about 16, the population value, and their standard deviation (that is, standard error) is about $5/\sqrt{2} = 3.54$.

Once you get the hang of how the simulation works, you can change these factors and parameters:

- The size of the samples (select from the $N=$ dropdown list in the third panel). The sample size choices are 2, 5, 10, 16, 20, and 25.
- Population characteristics control: To change the shape of the population distribution, use the dropdown box in the first panel. To change the mean of the distribution, use the cursor, as indicated on the graph.
- Number of repetitions:
  - For 1 sample at a time, use "Animated."
  - For 5 samples at a time, click the "5" button.
  - For 1,000 samples at a time, click the "1,000" button.
  - For 10,000 samples at a time, click the "10,000" button.

Follow these instructions carefully and work slowly. We guarantee you will learn a lot! Keep in mind that you are drawing random samples. The very idea of sampling means that your results will not exactly equal anyone else's.

1. Go to the "Sampling Distributions" page at the Rice Virtual Lab in Statistics (http://onlinestatbook .com/ statsim/sampling_dist/index.html).

2. Click "Begin" in the left-hand frame.

3. The Java applet will start.

4. Ignore unfamiliar statistics (for example, "kurtosis").

5. Make sure the population distribution is normal. Observe the population mean and standard deviation.

6. Go to the third panel and select "Mean" if it is not already set and choose $N=2$ for the sample size.

7. Click on the "Animated" button. (It's on the right side next to the "Sample Data" graph.)

8. You will see two "observations" drop into the "Sample Data" graph and the mean of the two observations drop onto the "Sampling Distribution of Means, $N=2$" graph. This is the first mean.
   a. What is the mean? (Look on the left side of the "Sample Data" graph.)

9. Click "Animated" again to draw a second sample of size 2. You will see the observed values drop down and then the mean of *this* second sample will appear in the sampling distribution panel. At this point you will have two sample means plotted. You can see their average on the far left-hand side of the page. Look in the third panel labeled "Distribution of means, $N=2$." You will also see the standard deviation, denoted "sd." This the standard error of the mean ($\hat{\sigma}_{\bar{Y}}$). As you proceed, the distribution of your sample means will be plotted and their average appears on the left.

10. Answer these questions:
    a. What is the average of the two? (Be sure to use the numbers from the third panel.) _____
    b. What is the standard error? (It's denoted "sd" in the third panel.) _____

11. Click "Animated" again.

12. And again and again until you have a total of ten replications (marked "Reps" on the page).
    a. What is the mean of the 10 sample means? _____
    b. What is the standard error? (*Note:* It will be less than the population standard deviation because the standard error equals $\sigma/\sqrt{N}$, where σ is the population standard deviation.) _____

13. Make a note of the results. With samples of size 2, any particular sample mean may be quite far from the population mean (16), but the more samples you draw, the closer the mean of the means gets to that true value.

14. Click "Clear lower 3" to reset the simulation.

15. Leave the sample size at 2 but click on the "1,000" button under "Sample" in the second level. In effect, you have taken 1,000 independent samples of $N = 2$ each from the same population, which has a bell-shaped (normal) distribution with a mean of $\mu = 16$ and a standard deviation of $\sigma = 5$. A graph of these means appears in the third panel. This is a picture of the sampling distribution, and its summary statistics appear on the left. You should observe that the mean of the 1,000 means is about 16.
    a. What is the mean of the sampling distribution? _____
    b. What is the standard error? (Remember: it will be about $\sigma/\sqrt{N}$, so it should be smaller than the population standard deviation.) _____

16. Now clear the simulation and obtain 10,000 samples. Note the mean of these 10,000 means.
    What is the mean? _____
    What is the standard error? _____
    What is the shape of the sampling distribution? _____

17. Now clear the simulation and set the sample size to $N = 20$. Repeat the simulation with first 1,000 replications and then 10,000. (Remember to reset by clicking the "Clear lower 3" button.)
    a. What is the mean of the 1,000 sample means (each based on twenty cases)? _____
    b. What would you expect the standard error (the standard deviation of the distribution of means, "sd") to be? _____
    c. What is the standard error ("sd") of these 1,000 sample means? _____
    d. What is the mean of 10,000 sample means? _____
    e. What is the standard error? _____

18. Time now to change the population distribution. Clear the simulation (that is, "Clear lower 3") and select "Uniform" from the dropdown list under the "Clear lower 3" box. Note the new population mean and standard deviation; they should be 16 and 9.52.

19. Click "1,000" to draw 1,000 samples of size 20 from a uniformly (rectangular) distributed population with mean 16.
    a. What is the mean of the 1,000 sample means drawn from this population? _____
    b. What is the standard error? _____
    c. What does the sampling distribution look like? _____

20. Clear the simulation and click on the "10,000" button.
    a. What is the mean of the 10,000 sample means drawn from this population? _____
    b. What is the standard error? _____
    c. What does the sampling distribution look like? _____

21. What general lessons do you draw from this exercise?

_____

_____

_____

_____

CHAPTER 8

Making Empirical Observations
Direct and Indirect Observation

Observation of political activities and behaviors is a data collection method that can be used profitably by political scientists. Although some observation studies require a considerable amount of time and researchers' presence in particular locations, most of us are likely to have used this method to learn about politics in a casual manner, without traveling great distances or staying in a setting for a long time. With some imagination, you can find numerous opportunities to use observation systematically to collect information about political phenomena. For example, you can judge community concern about proposed school budgets by attending school board meetings. You can observe the nature of political comments made by those around you and how others react. You can assess power relationships or leadership styles by observing physical and verbal cues given by participants in various settings.

Exercise 8–1. David A. Bositis notes in his article, "Some Observations on the Participant Method," *Political Behavior* 10 (Winter 1988): 333–48, that "a key feature of participant observation design is an ability to both observe behavior and to provoke behaviors to be subsequently observed." Read Bositis's article and think of situations in which your participation (either your physical presence or verbal communications) could provoke behaviors to be observed. Are any ethical considerations raised by these situations? If not, think of a situation that poses some ethical issues. If all of the situations you thought of pose ethical issues, try to think of one that does not raise ethical issues.

Exercise 8–2. Read Laura Beth Nielsen's article, "Situating Legal Consciousness: Experiences and Attitudes of Ordinary Citizens about Law and Street Harassment," *Law and Society Review* 34, no. 4 (2000): 1055–90, especially the "Method" section beginning on page 1061. (You can find the excerpt here in the workbook in chapter 7, exercise 7–1.) How did observation play a role in her research?

_____

_____

_____

_____

_____

_____

_____

_____

Exercise 8–3. Read James M. Glaser's article, "The Challenges of Campaign Watching: Seven Lessons of Participant-Observation Research," *PS: Political Science and Politics* 29 (September 1996): 533–37. (The article is available in JSTOR.) Why is participant observation important to studying political campaigns? How important is flexibility in this type of research?

_____

_____

_____

_____

_____

_____

_____

_____

Exercise 8–4. Here are some ideas for collecting data and testing propositions with direct and indirect measurement. This list is suggestive. Obviously, you will have to modify the topics to suit your needs and interests.

a. Formulate a hypothesis about the behavior of members of your city council or state legislature. (*Examples:* That in contentious debates Republican members appeal to patriotism more than Democrats, or that

Be Careful When Observing the Public

In chapter 8 we discuss several ways to make direct and indirect observations. In general, and especially if you are working through a university or college, you must obtain the informed consent of individuals you question in a poll or survey or use in an experiment. Getting this agreement may be a straightforward matter of asking for permission, which subjects should feel completely free to give or deny. You may, however, be involved in direct or indirect observation of people (or their possessions) that does not involve face-to-face contact. (Suppose, for example, you want to observe a protest march.) Even in this case you should accept some standards of responsible and courteous research:

■ Be aware of your personal safety. Make sure someone knows where you are going. Carry proper identification. It also wouldn't hurt to carry a letter of introduction from your professor, supervisor, or employer.

■ Depending on the nature of the study, contact local authorities to tell them that you will be in a certain area collecting data in a particular way. If you seem to be just "hanging around" a neighborhood or park or schoolyard, you are inevitably going to be reported as "someone acting suspicious."

■ Always ask permission if you enter private property. If no one is available to give it, come back later or try somewhere else. Even in many public accommodations such as arenas or department stores you will probably need to get prior approval to do your research.

■ Respect people's privacy even when they are in public places.

■ Do not misrepresent yourself. Here's what happened once to some of our students. They wanted to compare the treatment that whites and nonwhites received in rural welfare offices. But when applying for public assistance to observe the behavior of welfare officers by pretending to be needy they were breaking state and federal laws. They got off with a warning, but it's always a big mistake to fake being someone you're not just to collect data.

■ Be willing, even eager, to share your results with those who have asked about your activities. Volunteer to send them a copy of your study. (Doing so will encourage cooperation.)

■ When observing a demonstration, protest march, debate, or similar confrontation, do not appear to take sides.

women have different "political styles" than men.) After developing measures or indicators of this behavior, attend a public meeting or two (in some places local and state meetings or hearings are televised) to see if the hypothesis is supported. What did you find?

_____

_____

_____

_____

_____

_____

_____

_____

_____

_____

b. Think of a similar hypothesis for members of Congress and observe the floor or committee action of the Senate or House on C-SPAN. Can you, for instance, show a systematic difference in the behavior of different party members or representatives from different regions of the country? Are the debates more courteous in one chamber than in the other?

_____

_____

_____

_____

_____

_____

_____

_____

_____

c. It has been said that cars in the United States are a reflection of the owners' personalities, values, and status. Car owners often take a public stance on a controversial issue by displaying a politically based bumper sticker. Visit a location such as the parking lot of a university, a stadium, an airport, or a shopping mall. Pick a starting location, and for the first fifty vehicles, write down the make and model of the vehicle, whether or not the vehicle displays a political bumper sticker, and the nature of the message of any sticker you find. If a vehicle has more than one sticker, record the number of stickers and the nature of each message. Is there a correlation between make and model and whether the vehicle has a bumper sticker? Is there a correlation between make and model and owner's stand on issues?

_____

_____

_____

_____

_____

_____

_____

_____

_____

d. Visit the parking areas of several different places of worship and record the content of the bumper stickers on cars. Is there a difference in the nature of the messages displayed on the cars in the place of worship? Is there a difference in the nature of the messages displayed on the cars at different places of worship? Did you expect any differences?

_____

_____

_____

_____

_____

_____

_____

_____

_____

CHAPTER 9

Document Analysis
Using the Written Record

In chapter 9 we discuss using the record-keeping activities of institutions, organizations, and individuals as sources of data for research projects. In some situations, as with the *Statistical Abstract of the United States* or the World Bank's *World Development Indicators,* the records provide data that are directly usable as operational measures of concepts and variables, although you will need to decide how well these data measure the concepts you want to measure. In other cases, to create measures of your variables, you will need to analyze the records, using content analysis, for example. Much of the data you collect will be part of the "running" record, that is, collected on a routine basis by public organizations like the U.S. Census Bureau. It is less likely that you will use "episodic" records, that is, those records that are preserved in a casual, personal, and accidental manner.

Exercise 9–1. Access the *Digest of Education Statistics* at http://nces.ed.gov/programs/digest. Under "List of Tables and Figures," choose "2006 full version of the digest." From the list that pops up in the left navigation bar, click on the link that says "List of Tables by Chapter" and then choose "Opinions on Education" from the menu under "Chapter 1" in the center of the page. Next, click on table 22, which contains data on the average grade the public would give public schools in their community and the nation at large. (*Note:* These data can be downloaded into an Excel file.)

a. According to the data for 2006, do parents of children in public schools differ from adults who have no children in school with respect to their opinions about their own communities' schools? How about nationally?

_____

_____

_____

_____

b. For all adults, has the average grade given to public schools in the community and for the nation at large changed much between 1981 and 2006?

_____

_____

_____

_____

c. Return to the page for "2006 full version of the digest" at http://nces.ed.gov/programs/digest/d06, and from the left navigation bar choose "List of Figures." Click on chapter 5, "Outcomes of Educations," and click on figure 24, "Median annual income of persons 25 years and over, by highest level of education and sex 2005" and fill in the following table:

	Median Income	
Highest Level of Education	Male	Female
Some high school		
High school completion		
Associate's degree		
Bachelor's degree		
Master's degree		

d. Go back to the list of figures at http://nces.ed.gov/programs/digest/d06/figures.asp. Choose chapter 6, "International Comparisons of Education," and then find figure 27. Which country has the highest percentage of the population of typical age of graduation with bachelor's degrees, 2005? What is the percentage? What is the percentage for the United States?

_____

_____

e. From the top of the Web page, click on "Surveys and Programs." Scan down the various programs and view the various "lists of surveys" for the programs. You can find these at the bottom of each section. In particular, click on "Lists of Elementary/Secondary Surveys." Scan down and click on "Education Finance Statistics Center," and then click on "Finance Graphs." Click on graph 2, "Current per-pupil expenditures for elementary and secondary education in the United States: Fiscal year 2005." Which two states spent the most per pupil for elementary and secondary education for fiscal year 2005? Which state spent the least? Click on "View Data Table" to find out how much these states spent. How much did your state spend in 2005?

_____

_____

_____

Exercise 9–2. The World Bank is a source of a vast amount of statistics, particularly on factors related to economic development in countries around the world. These data and many reports may be accessed on its Web site, www.worldbank.org. From the home page, click on "Data" at the top of the page. Under "Data Catalog," click on "World Development Indicators." This will bring up the report *World Development Indicators 2011*. Click on "Part 2" and read the section on Primary Data Documentation on page 393 (listed as 206–7 in the reader).

a. What are some of the problems or sources of error in the data?

_____

_____

b. Why is the demand for good-quality statistics increasing?

_____

_____

Exercise 9–3. State of the State addresses are examples of primary documents that can be analyzed to measure the political climate and important policy issues confronting the states. Use the following link to access State of the State addresses given by state governors since 2000: http://www.stateline.org/live/resources/ Speech+Archives. You may want to scan a few of these before you perform the following exercises. Describe the procedures you might use to analyze the content of the speeches to measure the following:

a. The policy issues important to the governor in a particular state

_____

_____

b. The relative weight given to policy issues versus symbolic or rhetorical statements in a governor's speech

_____

_____

c. Whether a difference exists in the issues mentioned by Republican and Democratic governors

_____

_____

d. Now select four addresses and apply your measurement scheme. What did you find? Did you need to make changes to your procedures, or do you think they could be improved? Why?

_____

_____

_____

_____

_____

_____

_____

_____

_____

_____

_____

CHAPTER 10

Survey Research and Interviewing

The use of survey research and elite interviewing has become far more than a staple of empirical political science. These tools now guide decision makers' thinking about public policy, appear as data in support of partisan arguments, and offer the mass media a way to inform consumers about what is going on in the world. The reason is obvious: interviewing and polling supposedly provide objective "scientific" data. If, for example, a politician or an organization can claim that more than half of the public favors a particular position, that stance might acquire a legitimacy it would not otherwise enjoy. So widespread is the use of polls to acquire information about what people think that they have become part of everyday parlance of even apolitical citizens.

But as common and impressive as these methods are, they nevertheless need to be considered carefully. After all, everyone knows the adage, "Ask a dumb question, get a dumb answer." It behooves students of politics to become familiar with what the techniques can and cannot accomplish.

Just as in the natural sciences, measuring and recording devices in the social sciences rest on theories. In surveying and elite interviewing, the most fundamental premises are that respondents have certain information sought by the investigator, that they know how and are willing to provide it, and, perhaps most important, that everyone involved in the process shares to a high degree of approximation the meanings of the words and symbols employed in the information exchange. That is why chapter 10 stresses the importance of thinking clearly about how questions are worded and presented to respondents, who may or may not care much about the researcher's topic. We note in particular that if questions are ambiguous or threatening, people may not answer them or not answer them truthfully.

Please take some time to consider the assignments. They demand not only that you understand specific terms ("open-ended," for example) but also that you fully grasp the difficulty of designing an effective questionnaire, whether intended for an elite or a mass audience.

Exercise 10–1.   Suppose you are thinking of surveying the general public about "party identification," the psychological feeling of closeness or attachment to a political party. (Party identification is *not* party registration, which is a legal standing. It is an attitude or disposition toward a party.)

a.   Write an example of a closed-ended question.

_____

_____

_____

_____

b. Write an example of an open-ended question.

_____

_____

_____

_____

c. Write an example of a reactive question.

_____

_____

_____

_____

d. Write a question that measures the *strength* or *intensity* of party identification. (To see how major research organizations phrase such a question, look on the Internet, for example, for "The ANES Guide to Public Opinion and Electoral Behavior" at www.electionstudies.org/nesguide/nesguide.htm.)

_____

_____

_____

_____

Exercise 10–2.   For each of the concepts or topics listed following, which would be most appropriate, a closed- or an open-ended question? Why? If either would be appropriate or possible, explain in sufficient detail to demonstrate your knowledge of the difference between the types of questions.

a. Whether or not a person has lived in the congressional district for more than two years

_____

_____

_____

_____

b. Actions respondent is willing to take to reduce greenhouse gas emissions

_____

_____

_____

_____

c. A person's support or opposition to embryonic stem cell research

_____

_____

_____

_____

d. A person's satisfaction with the current Congress's performance

_____

_____

_____

_____

e. How many hours a week a person works

_____

_____

_____

_____

f. What a person likes or dislikes about Representative Michele Bachmann

_____

_____

_____

_____

g. A person's rating of Senator Harry Reid's performance in office

_____

_____

_____

_____

h. Support for restructuring Medicare

_____

_____

_____

_____

Exercise 10–3.   Provide a short evaluation or critique of the following survey questions. If you do not see any problems, just say so.

a.   "What is your opinion of a national flat tax? Do you oppose or favor it, or don't you have an opinion?"

_____

_____

_____

_____

b.   "When talking with citizens we find that most of them oppose increasing the federal tax on gasoline. How about you? Would you favor or oppose an increase in the gasoline tax?"

_____

_____

_____

_____

c.   "Would you favor or oppose the state selling its bonds and securities to private companies, if it would raise monies to pay for child health care?"

_____

_____

_____

_____

d.   "Since the September 11, 2001, attacks in New York City and Washington, D.C., there has been a great deal of attention paid to terrorism. Do you support increased spending to protect the nation from terrorism?"

_____

_____

_____

_____

e.   "Have you ever used an illegal drug?"

_____

_____

_____

_____

Exercise 10–4. Sometimes it is difficult to write high-quality survey questions that are both reliable and valid (see chapters 5 and 10). It is particularly important to consider content validity (whether you have fully captured the meaning of a concept with a survey question). In this exercise you are to, first, provide a definition for each of the terms listed, second, write a survey question for each, and finally, explain how each question captures the full meaning of the term.

a. Party identification

_____

_____

_____

_____

b. Attitude toward female office holders

_____

_____

_____

_____

c. Level of support for the British prime minister

_____

_____

_____

_____

d. Ideology

_____

_____

_____

_____

e. Level of participation in politics

_____

_____

_____

Exercise 10–5. You can explore the effects of question wording on the distribution of responses and hence on an overall assessment of public opinion by examining polls that ask questions in different ways.

1. Go to the Survey Documentation and Analysis site (http://sda.berkeley.edu).

2. Click on "SDA Archive."

3. Click on "GSS Cumulative Datafile 1972–2010."

4. A page titled "SDA Web Application" will appear. At the top of the page find a blue bar with, among other items, "Codebooks" on it. (It is the fourth blue box from the left.) Place the cursor over "Codebooks" and choose "Codebook by Year of Interview" from the dropdown menu and click.

5. A new page appears. On the left, find and choose "Alphabetical Variable List."

6. On the next page choose "Go to page containing range of items: POLEFF4 ZOMBIES." (It's the third or last row.)

7. Scroll down to and click "Trust." The question is, "Can people be trusted?"

8. The next page contains responses to the question, "Generally speaking, would you say that most people can be trusted or that you can't be too careful in life?"

9. Scroll down to 1983 to find the responses given in 1983.
   a. What percentage (frequency) chose "Can trust"? _____
   b. What percentage (frequency) chose "Cannot trust"? _____
   c. What percentage (frequency) chose "Depends"? _____

10. Go back one screen to the alphabetical list of variables and scroll down to "TrustY." The question is, "Do you think most people can be trusted?"
    a. What percentage (frequency) chose "Yes"? _____
    b. What percentage (frequency) chose "No"? _____

11. Discuss your findings. Remember the questionnaire was administered by a reputable scholarly institution to respondents in the same year, 1983. If you find a difference in response patterns, what do you think explains it?

_____

_____

_____

_____

Exercise 10–6. Continue using the GSS study to explore the effects of variations on question wording on issues involving any or all of these topics: abortion (start by examining variables "Abany" to "Abhave3"), gun control ("Crimdown" and "Crimup"), legalization of marijuana ("Grass" and "GrassY"), cancer screening for employment ("GenecanX" and "GenecanY"), attitudes toward government spending on the environment ("Natenvir" and "Natenviy"), and improving the condition of African Americans ("Natrace" and "Natracey"). There are no set answers here. Sometimes question wording makes a difference, sometimes it doesn't. What is important is that you spend time thinking about how phrasing, the availability of options, "code" or inflammatory words, and so on might affect responses.

Exercise 10–7. Imagine that you are a student who is bored with an assignment on survey question writing. The assignment asks you to write a series of five questions you would ask of voters in a presidential election. To make things more interesting you decide to write five questions that break all of the rules of question writing. In the space provided write five bad survey questions and explain why they break the rules.

1. _____

   _____

   _____

   _____

2. _____

   _____

   _____

   _____

3. _____

   _____

   _____

   _____

4. _____

   _____

   _____

   _____

5. _____

   _____

   _____

   _____

CHAPTER 11

Making Sense of Data
First Steps

Chapter 11 begins the study of applied statistical analysis. Its main goal is to introduce in a nontechnical, non-threatening way some tools that can be used to summarize a batch of numbers and make inferences. Besides being part and parcel of all fields of political science, many of these concepts appear in the mass media. Furthermore, political science leads to quite a few interesting and exciting career opportunities, and most of them require at least a rudimentary knowledge of quantitative research methods. Think, for example, of someone playing a major role in an election campaign. In all likelihood, she or he will have to analyze poll data or at least interpret and critique someone else's analysis. Or suppose you have an internship in a government agency. You may be of greater assistance to your employers if you can provide a modest amount of technical advice about reports they are receiving or advice the agency supplies the public. So there are lots of reasons for studying at least a few quantitative methods, no matter how far removed from the world of politics they seem to be.

Many students are initially put off by having to learn statistics, but our experience tells us that this aversion often results from unfamiliarity with the subject, not its inherent difficulty. So even if you are one who says, "I stink at math," at least attempt to keep an open mind. We think you may find that these concerns are misplaced.

Here are a few tips:

- **Keep up.** Unlike some subjects that may seem to lend themselves to cramming, statistics is best learned step by step; you should make sure you understand each concept reasonably well before moving on to the next one. And since the ideas are possibly daunting at first sight, it is easy to get lost if you try to learn everything all at once. This is, in short, one course where it pays to stay on top of the readings and assignments.

- **Learn by doing.** You can't get into good physical shape by reading articles on conditioning. You have to work out regularly. In the same way and for essentially the same reasons, data analysis has to be learned actively. It is crucial that you perform your own analysis. Simply reading about how it is done will not give you the functional understanding that makes statistics so useful. The exercises in this workbook are designed to do just that: give you actual training in data analysis.

- **Keep substance over method**. Whenever possible think about the substantive context of a problem. You may be asked to calculate a mean or standard deviation, for example. But what is important is not the numbers per se (although they do have to be correct), but rather what they say about the problem at hand. For example, instead of just writing, "The average is 10," you should write, "The average is 10 thousand dollars," to keep firmly in mind that you are working on a concrete issue and not an abstract algebra problem.

- **Be neat and orderly.** Yes, this advice sounds peevish. Yet we have found that a huge number of mistakes and misconceptions arise simply from disorderly note taking and hand calculations. It is always a good idea to have plenty of scrap paper handy and to work in a top-down fashion rather than jump all around the page putting intermediate calculations here and there in no logical order. It should be possible for you or anyone else to reconstruct your thought processes by following your calculations from beginning to end. That way errors and misunderstandings can be spotted and corrected.

■ **Don't trust the computer.** Many of the questions we ask can be answered only with the assistance of a computer. And we are the first to admit that computers are marvelous devices. But they have no ability to grasp what you mean to type and do not have the common sense to decipher what to you may be obvious. (We have stressed in the text on several occasions that a computer can only do what it is told and consequently will not make a mistake unless directed to do so.) So every time you turn on the computer, be prepared. Ahead of time ask yourself, "What do I need to find out? What procedure will give me the answers?" If you find yourself getting frustrated or that something does not work no matter how many times you try, back off. Turn off the machine, go for a brief walk, rewrite your questions on a fresh piece of paper, and then go back to the system.

In these assignments we ask you to examine one variable at a time. The idea is to summarize a possibly large batch of numbers with a few indicators of a distribution's central tendency, variation, and shape. We have also appended a brief set of guidelines for preparing your own data for analysis.

Exercise 11–1. Suppose you want to write a term paper titled "The Causes of Crime in America." Among other research hypotheses you are considering is the idea that urban crime rates will at least be associated with demographic factors such as education, poverty, and social-ethnic heterogeneity. You start collecting data by consulting *State and Metropolitan Area Data Book: 2006*, a regular publication of the US Census Bureau.[1] Figure 11–1 shows a small excerpt from table B-4 of this source.

FIGURE 11–1
Raw Data

Table B-4. Metropolitan Areas—Population Characteristics 2000 Census

Metropolitan statistical area with metropolitan divisions *Metropolitan division*	Households Total	Family with children under 18 years (percent of total)	High school graduate or higher (percent)	Bachelor's degree or higher (percent)	Foreign-born population (percent of total)	Speaking language other than English at home[2] (percent)	Living in same house in 1995 and 2000[2] (percent)	Workers who drove alone to work[3] (percent)	Households with income of $75,000 or more in 1999 (percent)	Persons below poverty in 1999 level (percent)
Abilene, TX	58,475	37.7	78.7	19.6	3.5	13.8	47.6	80.8	12.6	14.6
Akron, OH	274,237	33.6	85.7	24.3	3.0	5.3	57.8	85.4	21.7	9.6
Albany, GA	57,403	40.7	73.2	15.7	1.6	3.8	55.2	79.4	15.0	20.9
Albany-Schenectady-Troy, NY	330,246	32.5	86.1	29.1	4.8	8.1	59.3	79.6	23.1	9.0
Albuquerque, NM	281,052	36.6	83.7	28.1	7.8	30.1	50.6	77.5	19.0	13.7
Alexandria, LA	54,193	39.4	74.4	15.7	1.4	4.6	60.2	78.6	12.1	20.0
Allentown-Bethlehem-Easton, PA-NJ	285,808	33.5	NA	21.7	5.2	12.0	60.9	81.9	23.2	8.0
Altoona, PA	51,518	31.9	83.8	13.9	1.0	3.1	66.7	82.2	10.8	12.3
Amarillo, TX	85,272	37.8	80.1	20.8	6.0	16.4	48.7	82.0	15.5	13.0
Ames, IA	29,383	28.2	93.5	44.5	6.9	9.4	39.3	71.3	19.9	12.4
Anchorage, AK	115,378	42.2	89.9	27.0	7.1	12.0	43.1	73.5	32.1	7.9
Anderson, IN	53,052	33.1	80.1	14.4	1.2	3.1	59.7	81.8	17.4	9.0

Remainder of table not included.

NA—not available.

Source: US Census Bureau, *State and Metropolitan Area Data Book: 2006.*

[1] US Census Bureau, *State and Metropolitan Area Data Book: 2006.* Available at http://www.census.gov/compendia/databooks/pdfversion .html.

In the space below provide a data matrix that contains the name of the county, the percentage of residents who have graduated from high school, the percentage who are foreign born, and the percentage living in poverty in 1999. Remember: the point of creating a data matrix is to organize text and numbers so that they can be processed by hand or computer. Hence, be neat! Your table should also include the source of the data and how you have treated any missing data. (Use table 11–1 of the main text as a guideline.)

Exercise 11–2. Here are some variables that might arise in political research. For each one, what would be an appropriate measure of central tendency: the mean, the mode, or the median? If more than one seems useful, explain why.

a. The percentage of the population living in a metropolitan area who live below the poverty line

_____

_____

b. The per capita incomes of the nations of Europe and Africa

_____

_____

c. A four-point scale of attitudes toward immigrants

_____

_____

d.   The amount (in thousands of dollars) that a random sample of 800 New Yorkers contributed to the two major political parties in 2010

_____

_____

e.   The number of civil disturbances in 2011 in a sample of 24 African and Latin American nations

_____

_____

f.   The number of times Supreme Court justices voted to overturn an act of Congress during the past forty years

_____

_____

g.   The number of toxic waste sites in each county in Pennsylvania

_____

_____

Exercise 11–3.   Table 11–1 displays some poll results as they were presented in the *Washington Post*.[2] This *Washington Post*-ABC News poll was conducted by telephone July 14 to 17, 2011, among a random national sample of 1,001 adults, including users of both conventional and cellular phones. The results from the full survey have a margin of sampling error of 3.5 percentage points.

What information is missing from this table that might be important in interpreting the results?

_____

_____

TABLE 11-1

Poll Question

Who is not willing enough to compromise on the budget deficit?

	1/04
Republican Leaders	77%
Obama	58%

---

[2] Dan Balz and Jan Cohen, "Post-ABC Poll: GOP Too Dug In on Debt Talks; Public Fears Default Consequences," *Washington Post*, July 19, 2011. Available at http://www.washingtonpost.com/politics/post-abc-news-poll-public-sees-dire-consequences-if-no-budget-deal/2011/07/19/gIQA4MQPOIstory.html?hpid=z1.

Exercise 11–4. The 2008 National Election Study posed the following question: "Some people believe that we should spend much less money for defense. Others feel that defense spending should be greatly increased. Where would you place yourself on this scale or haven't you thought much about this?"[3] Here is a tally of responses to the question:

Greatly decrease	1:	82
	2:	82
	3:	128
	4:	280
	5:	210
	6:	128
Greatly increase	7:	93
Don't Know:		164

a. In the space following create a frequency distribution for the responses that includes the raw numbers, the relative frequencies (or proportions or percentages), the "valid" relative frequencies (or proportions or percentages), and the cumulative proportions or percentages. (Treat "don't know" as missing data.)

b. What is the modal category?

_____

_____

Exercise 11–5. The data sets available on the Web site at http://psrm.cqpress.com include two versions of a very small portion of the data from the "British Election Study 2001/02." One is in an SPSS portable document format, "bes2001.por," and the other in a plain text format, "bes2001.dat." The description of the variables is in a file called "bescodebook.txt," which is a plain text file that can be opened with just about any word processor. Perform the following analyses using the software your instructor requires:

---

[3] The American National Election Studies (http://www.electionstudies.org). *The 2008 National Election Study* [data set]. Ann Arbor: University of Michigan, Center for Political Studies [producer and distributor].

a. In the space following create frequency distributions for "Your financial situation" ("Economic evaluation—self") and "the Nation's financial situation" ("Economic evaluation—country").

b. Overall, how do the distributions compare? What is the modal response in each frequency distribution? What substantive conclusions do you reach?

_____

_____

_____

_____

_____

_____

c. Using the same general technique we used in chapter 11 to compare one group with another, compare men's and women's attitudes toward British membership in the European Union (EU). (*Hint:* What is the independent variable? Its categories go where on the table: along the side or at the top? What is the dependent variable and where do its labels appear? What percentages should your software compute?)

_____

_____

_____

_____

_____

_____

Exercise 11–6. Using the hypothetical data in table 11–2, calculate the sample variance and sample standard deviation. Next, calculate the population variance and the population standard deviation using the same data.

After completing your calculations, explain in plain English why the answers are different for the sample and population equations. How and why are the formulae different?

_____

_____

_____

_____

_____

_____

TABLE 11–2
Political Events Attended

Respondent Identification	Number of Political Events Attended in Last Year
1	4
2	3
3	0
4	1
5	7
6	3
7	1
8	0
9	5
10	1

Exercise 11–7. Table 11–3 contains some data that might come up in a discussion of gun control. (*Note:* Your instructor may permit you to use statistical software to find or check the answers, but the requested information can be obtained with a decent calculator.)

a. What is the mean death rate from firearms?

_____

b. What is the median death rate?

_____

TABLE 11–3
Death Rates from Firearms Injuries, Selected Countries

Country	Death Rate from Firearms[a]
United States	13.7
Australia	2.9
Canada	3.9
Denmark	2.1
England and Wales	0.4
France	6.3
Israel	2.8
Netherlands	0.5
New Zealand	3.1
Norway	4.3
Scotland	0.6

Source: US Bureau of the Census, *Statistical Abstract of the United States: 1999,* 119th ed. (Washington, DC 1999), 837.

[a] Deaths per 100,000 population.

c. What is the trimmed mean? (Assume one value from each end of the distribution has been cut.)

_____

d. Comment on the difference between the median (or trimmed mean) and the mean in this context.

_____

_____

_____

e. What is the maximum rate?

_____

f. The minimum?

_____

g. What is the first quartile (Q1)? _____ The third quartile (Q3)? _____

The interquartile range (IQR)? _____

h. What is the simple sum of the deviations from the mean? (*Hint:* You don't need a calculator for this one.)

_____

i. What is the sum of the absolute values of the deviations?

_____

j. What is the mean deviation?

_____

k. What is the total sum of squares (TSS)?

_____

l. What is the sample standard deviation?

_____

m. What is the sample variance of the death rates?

_____

n. Draw a box plot to describe the shape of the distribution. (*Hint:* You can determine this by using some of the quantities calculated in the preceding questions.)

o. How would you describe the distribution's shape?

_____

_____

_____

p. Draw a dot plot for the firearms data.

q. Drop the largest and smallest observations from this data set, leaving nine observations. What is the new mean? The new median?

_____

_____

r. What is the new standard deviation? How does it compare to the one based on all eleven cases? If there is a difference, explain the cause.

_____

_____

_____

ONLINE STATISTICAL PROGRAMS

One of the benefits of the Internet is the wide availability of statistical programs. Many allow you to enter small amounts of data such as are found in these exercises. The programs vary greatly in content and quality, but you can use them to obtain many introductory statistical analyses. A couple of examples:

■ "One Variable Statistical Calculator" (http://bcs.whfreeman.com/ips4e/cat_010/applets/histogramIPS.html): Results include mean, standard deviation, and location statistics (for example, minimum, maximum, IQR, median), and histogram.
■ Larry Green's Applet Page, "Calculating One Variable Statistics" (www.ltcconline.net/greenl/java/index.html#Computation): Provides the mean, median, mode, maximum, minimum, range, variance, and standard deviation.
■ "Statistical Applets" (www.assumption.edu/users/avadum/applets/applets.html): Calculates mean, standard deviation (calculated two ways), variance, sum of squares, median, minimum, maximum, 25th percentile, and 75th percentile.

As abundant and easy to use as these resources are, pay attention to these potential problems:

■ Your instructor may want you to use a single program. There is a good reason for this requirement: not all software computes statistics in the same way. Your answers may be valid according to one program but not another. Trying to sort these matters out can be extremely difficult.
■ As the text alludes to in several places and as we mentioned earlier, some statistics and most graphs can be calculated correctly in different ways. For example, the text obtains the standard deviation by obtaining the square root of the sum of squares divided by $N-1$. Many textbooks, however, instruct you to divide by simply N. If the number of cases is not large, the two versions of the standard deviation will differ. So if you use one program and someone else uses another, your results may not agree.
■ Most of the sites use Java applets, which means your browser has to support Java. Some applets take forever and a day to load, especially if the connection speed is slow.
■ Internet sites come and go. Make sure you do not need to return to one to finish an assignment.
■ Commercial sites may or may not let you use their software with registration and/or a fee.

In spite of these possible pitfalls, the Internet can be an excellent source for supplementing your introduction to applied statistics.

Exercise 11–8. Table 11–4 contains "abortion ratios" (that is, the number of abortions performed per 1,000 live births) for a random sample of eight American states.

Compute these summary statistics for the abortion data:

a. Mean: _____

b. Median: _____

c. Range: _____

d. Standard deviation: _____

e. What is the first quartile (Q1)? _____

The third quartile (Q3)? _____

The interquartile range (IQR)? _____

f. In the space following draw a box plot and a dot plot of these data.

TABLE 11–4

Abortion Ratios in Eight American States

State	Abortion Ratio
Delaware	368.8
Wyoming	1.0
Iowa	155.0
Florida	424.0
Illinois	231.5
North Dakota	169.8
Oklahoma	129.0
Connecticut	289.3

Source: Data are from "Abortion Statistics and Other Data," accessed October 27, 2005, from www.johnstonsarchive.net/policy/abortion/index .html#US. The "Abortion Statistics" data originally came from many public and nonprofit sources and were compiled by Wm. Robert Johnston.

g. Briefly describe the distribution.

_____

_____

_____

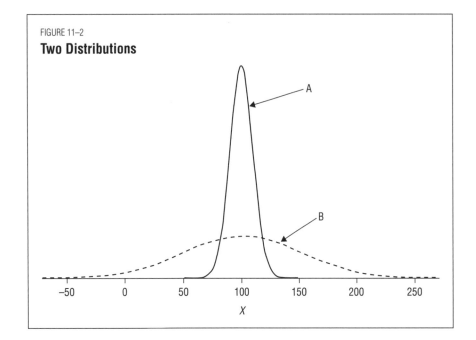

FIGURE 11–2
Two Distributions

Exercise 11–9. Figure 11–2 contains two distributions, A and B.

a. Which distribution has the greater variation? _____

b. Which distribution has the smaller standard deviation? _____

c. How do measures of central tendency for distributions A and B differ, if they differ at all?

_____

d. What are the mean and the median of distribution A? _____

e. What are the mean and the median of distribution B? _____

Exercise 11–10. During the 2004 presidential campaign there was a lot of discussion about who benefited from the tax cuts initiated and signed into law by the George W. Bush administration. Senator John Kerry said, "George Bush's only economic plan is lavish tax breaks for those at the top." President Bush asserted proudly, "I have twice led the United States Congress to pass historic tax relief for the American people." Part of the argument turned on dollar amounts received by different groups. A White House news release claimed, "Under the President's proposal to speed up tax relief, 92 million taxpayers would receive, on average, a tax cut of $1,083 in 2003."[4] Yet one of the president's critics wrote, "The average working family would get about $289."[5] Assuming both sides are telling the truth, how do you suppose they could reach such different conclusions? You will not be able to provide a definitive answer, but your knowledge of summary statistics should give you a good idea.

_____

_____

[4] "Fact Sheet: President Bush Taking Action to Strengthen America's Economy." Available at http://www.whitehouse.gov/ws/releases/2003/01/20030107.html.

[5] Kathryn Casa, "The Elephant in the Room," *CounterPunch*, edited by Alexander Cockburn and Jeffrey St. Clair. Available at http://www.counterpunch.org/casa01292003.html

_____

_____

_____

_____

_____

_____

_____

_____

Exercise 11–11. Define as clearly as possible the following statistics. Write a formula if you wish, but try to explain each term in plain language as well.

a. The sample standard deviation

_____

_____

_____

_____

_____

b. The median

_____

_____

_____

_____

_____

c. The mode

_____

_____

_____

_____

_____

d. The trimmed mean

_____

_____

_____

_____

_____

e. The interquartile range

_____

_____

_____

_____

_____

Exercise 11–12. Table 11–5 presents per capita expenditures by state and local governments for parks and police operations.[6]

TABLE 11-5

Per Capita Expenditures for Parks and Police Operations

Year	Current Expenditures for Parks and Recreation	Current Expenditures for Police	Year	Current Expenditures for Parks and Recreation	Current Expenditures for Police
1977	$48	$146	1992	$65	$184
1978	$50	$145	1993	$64	$181
1979	$49	$140	1994	$64	$185
1980	$47	$134	1995	$66	$188
1981	$47	$134	1996	$66	$195
1982	$48	$138	1997	$68	$201
1983	$50	$145	1998	$70	$208
1984	$51	$147	1999	$71	$213
1985	$53	$153	2000	$73	$218
1986	$56	$160	2001	$75	$220
1987	$58	$167	2002	$79	$232
1988	$60	$169	2003	$82	$236
1989	$61	$168	2004	$80	$236
1990	$63	$174	2005	$80	$242
1991	$64	$177			

Source: Tax Policy Center.

_____

[6] The data are available at the Tax Policy Center, State & Local Finance Data Query System. Available at http://www.taxpolicycenter.org/slf-dqs/pages.cfm.

a. Prepare a time-series plot of police and parks data on a single plot. (If no software is available, try to sketch the plot.) Submit the graph as your instructor requests.

b. Which trend appears to be growing fastest?

_____

Exercise 11–13. You can calculate percentage change in a trend by using this simple formula:

$$\text{Percentage Change} = \left(\frac{(\text{Last Number} - \text{First Number})}{\text{First Number}} \right) 100$$

Look at table 11–5 and answer the following questions.

a. What is the percentage change in expenditures for parks from 1977 to 2005? _____

b. What is the percentage change in police expenditures in the same period? _____

c. Do these calculations alter the judgment you made in exercise 11–12? How so?

_____

_____

_____

_____

_____

_____

APPENDIX TO CHAPTER 11

Preparing Data for Analysis

You may at some point be asked to collect data on your own. The text and the workbook provide examples of how the information can be organized for analysis by hand or computer. Here are a few specific tips that might speed up the process and help you avoid common mistakes.

Think of this step as more than one of organizing and cleaning the data so that they can be easily analyzed by hand or (more likely) computer. Indeed, the step has theoretical importance as well. Among other tasks, this process requires checking for and correcting errors, looking for inconsistencies (for example, a man who reports having had an abortion), recoding or changing recorded information to make it more analytically tractable, combing or separating categories of nominal variables, and determining what are to be considered valid responses. This step is often invisible because research organizations (such as the one that conducts the National Election Studies) do much of the work before releasing the information to others. But even so, end users have to know what has been done and frequently make additional adjustments such as deciding what to do with "don't know" versus "not interested" responses. Moreover, if you are collecting statistics from scratch, you have to put the information into a layout that facilitates tabulation and analysis. This process has important implications because at the end of the day these are the actual numbers used in constructing, testing, and modifying hypotheses and models. You would be amazed at how supposedly innocuous technical details can affect substantive conclusions.[7]

To appreciate this argument, let's use the topic of "prerequisites of democracy" as a concrete example. A general statement of one hypothesis is "The greater the degree of economic and social development, the more 'democratic' a nation will be." Suppose we have collected some data to investigate the hypothesis. Look at table 11–6, which shows data for a "sample" of thirty-six nations.[8] The table contains operational indicators of socioeconomic development and levels of democratization. (Table 11–7 in the text describes the variables in a bit more detail.)

Note first that the table comprises what we call a "data matrix," which, as we explained in the text, is a rectangular array in which the rows contain data for individual observations (one row per case) and columns contain variables. Austria, for instance, has an infant mortality rate of 4 deaths for every 1,000 live births; about 462 telephone lines for every 1,000 people; and so forth. The "level of democracy" variables are actually indices constructed from judgments of experts and are presented in various formats running from ordinal (for example, political rights) to interval (voice accountability) scales. (Akin to a spreadsheet, the table also corresponds to data storage in a computer.)

The table reveals some of the choices we made. Similar choices have to be made in nearly every study, although they may not be apparent.

- **Recording numbers.** Most computer programs insist that you enter numbers *without commas*. So, for example, 5,000 would be typed 5000. Moreover, *do not* use symbols such as the dollar ($) or percent (%) signs.
- **Plus and minus signs.** If the original data contain negative values, they have to be entered as such (see the "VA" column in table 11–1, for example). Positive values are always entered *without* a plus sign.
- **Precision.** We use at most two decimal places, mainly because the original sources contained only that many. Precision can be a tricky problem in statistical analysis. On one hand, in long, involved calculations so-called round-off errors (the rounding of decimal numbers during intermediate steps) can rapidly

---

[7] Clifford Clogg shows how the treatment of "don't know" responses to a question about the judicial system's toughness on criminals affects substantive conclusions. Clifford Clogg, "Using Association Models in Sociological Research: Some Examples," *American Journal of Sociology* 88 (July 1982): 114–34.

[8] The data in table 11–6 constitute a partial random and judgmental sample, not a purely random sample of the world's nation-states. For the sake of simplicity, however, we treat the numbers as if they were a truly random sample from an infinite population.

accumulate and lead to results that differ from those based on exact numbers. On the other hand, when you present data tables to others, one or two decimal places is usually sufficient.

- **Labels versus names.** To facilitate computer analysis, each variable has a short label or tag. Most computer programs require relatively brief names or abbreviations like the ones used here, but they can often handle separately entered longer descriptive labels. Note, however, that if a data matrix is to be published, every effort should be made to assign intelligible names to the variables; otherwise, readers can easily lose the meanings of the tags.

- **Category combinations.** Sometimes you have to combine categories to achieve an optimal number of cases. For example, we assigned New Zealand to Asia and Morocco and Egypt to the Middle East to avoid having categories with just one or two observations. The key is to make the assignments as explicit as possible so that others can understand and, if need be, challenge or modify them.

- **Recoding.** It may or may not be necessary, depending on software, to recode text to numeric values. The level of development ("status") for instance, could have been simply reported as "developed" or "developing," but we chose numeric codes 1 and 5, respectively, to stand for the substantive categories. The designation was arbitrary; we could have used 1 and 2 or 50 and 500 or any other two numbers. By the same token we could have (and still can) recode the region labels (for example, Afr, Asia) into numbers. Since development status and region are nominal variables, the numbers do not have an intrinsic quantitative meaning. But, surprisingly perhaps, it is legitimate to make use of these arbitrary numbers in some statistical procedures. We show how in chapter 13.

- **Missing data.** What do you do when data for a particular observation are not available because they were not collected or reported? This is the missing data problem. Look at table 11–6. Notice that no number appears in the "phone" variable for Luxembourg, Norway, and four other countries. Not having these data might encourage us to drop those units entirely from the analysis. Yet there is information for the remaining variables for those nations, so we decided to let an asterisk (*) stand in for the missing values. When our computer program encounters this symbol, it deletes the case for a particular procedure involving that variable and continues with its work. If we wanted the average of number of phone lines per 1,000 people across the entire sample, for example, the program would just skip Luxembourg and the others with asterisks in making the calculation. The result would be based not on all thirty-six countries but on the thirty with "valid" values. Do the countries with and without reported values differ in any systematic ways? It is hard to tell, which explains why a lot of thought has to go into collecting and preparing data for analysis. Each of these seemingly trivial matters can have significant implications.[9] A rule of thumb is that if 20 percent or more of your cases have missing values for a variable, you might consider dropping it or finding one that has more complete information.

- **Error and consistency checking.** Always check for errors and inconsistencies. In many instances, preliminary descriptive and exploratory analyses will alert you to the possibility of errors in the data. But this is not always the case, so if you are collecting data by hand, check and recheck the numbers.

- **Weights.** When certain sampling designs are used to collect the data, it may be necessary to adjust the numbers to reflect over- or under-representation of certain groups. This is not an issue in the comparative data we used to analyze levels of democratization, but many academic polls contain weighted data. The 2004 National Election Study does, and we have taken care of the weighting in our analyses. If you download data from the Internet, always check. (Most, but alas not all, software allows you to identify a weighted variable and will adjust the data for you.) In this book, however, weighting is never a problem.

---

[9] An enormous amount of thought has gone into the problem of missing data. For a review of a few statistical solutions, see Joseph L. Schafer and John W. Graham, "Missing Data: Our View of the State of the Art," *Psychological Methods* 7 (2002): 147–77.

CHAPTER 12

Statistical Inference

Chapter 12 expands on the introduction to statistics found in chapter 11, moving from basic statistical concepts like central tendency and dispersion to more complex hypothesis testing and confidence intervals. While chapter 11 explains how statistics can be used to explore data, chapter 12 focuses on making inferences using statistics. In this chapter in the workbook you are asked to use your new statistical skills to analyze data to answer questions.

Exercise 12–1. For this exercise you should refer to the definition and explanation of type I and type II errors in chapter 12. In the space provided, create an example involving a research assistant who commits a type I error. In the example, explain the consequences of making a type I error.

_____

_____

_____

_____

_____

_____

Exercise 12–2. Please review the binomial sampling distribution in table 12–2 in the text and the associated coin toss example in chapter 12. For this exercise you will put the example into practice by flipping a coin ten times and recording the number of heads you observe in row 1. You should then proceed to flip the coin ten more times and record your observations on each of the subsequent rows. Once you have recorded your observations on all ten rows, use a bar graph (review chapter 11) to visually display the results from the ten series (the number of heads observed in each iteration) in the space to the right. How does tossing the coins demonstrate the concept of the sampling distribution?

Series 1: ____ heads

Series 2: ____ heads

Series 3: ____ heads

Series 4: ____ heads

Series 5: ____ heads

Series 6: ____ heads

Series 7: ____ heads

Series 8: ____ heads

Series 9: ____ heads

Series 10: ____ heads

Exercise 12–3.   Many of the statistics in chapter 12 rely on the concept of statistical significance. To assert statistical significance one can compare a test or observed value to a critical value from a distribution like the normal or Student's t. In appendix B of the textbook you will find a table containing critical values from the t distribution. For each set of circumstances following indicate whether the observed value is sufficiently great to assert statistical significance.

1. $t_{obs} = 2.1$, 15 degrees of freedom, two-tailed test, 95% confidence level: _____

2. $t_{obs} = 1.9$, 24 degrees of freedom, two-tailed test, 90% confidence level: _____

3. $t_{obs} = 2.5$, 11 degrees of freedom, one-tailed test, 99% confidence level: _____

4. $t_{obs} = 1.5$, 30 degrees of freedom, one-tailed test, 95% confidence level: _____

5. $t_{obs} = 2.6$, 4 degrees of freedom, two-tailed test, 90% confidence level: _____

6. $t_{obs} = 2.01$, 50 degrees of freedom, two-tailed test, 95% confidence level: _____

7. $t_{obs} = 2.4$, 800 degrees of freedom, one-tailed test, 99% confidence level: _____

8. $t_{obs} = 3.5$, 19 degrees of freedom, two-tailed test, 99.8% confidence level: _____

9. $t_{obs} = 1.9$, 7 degrees of freedom, one-tailed test, 99% confidence level: _____

10. $t_{obs} = 4.0$, 21 degrees of freedom, two-tailed test, 99.9% confidence level: _____

Exercise 12–4.   The next set of questions is designed to help you get used to translating a political claim into a statistical hypothesis that can be tested.

According to the US Census Bureau, "The nation's public school districts spent an average of $8,701 per student on elementary and secondary education in fiscal year 2005, up 5 percent from the previous year."[1] A staff member for a candidate for governor has conducted a random sample of fifteen school districts and found that the mean spending level is only $8,000 per pupil. The candidate uses this finding to support his charge that the incumbent is weak on education. The newspaper you work for wants to know whether the difference between the population mean ($8,701) and the sample mean suggests that on the whole your state spends less on education than the rest of the country or results from sampling error.

a. Write a null hypothesis for this problem.

_____

_____

b. Write an alternative hypothesis. (*Hint:* Think carefully about the context. The candidate's argument is that the state spends less than . . . ?)

_____

_____

c. What statistical test would you use to evaluate the null hypothesis? Why?

_____

_____

_____

d. What would be the appropriate sampling distribution? Why?

_____

_____

_____

Exercise 12–5.   Here are three very easy questions. They are easy because there are (within broad limits) no right or wrong answers. Instead, your responses will be subjective; your opinion may differ from that of your classmates. The only point of the exercise is to encourage you to think about the costs of different types of errors when making inferences about unknowns. You might think of these as topics for discussion.

a. Imagine that you sat on a jury in which the defendant is accused of aggravated first-degree arson and murder. (In your state a crime of this severity always leads to the death penalty upon conviction.) Both the prosecution and the defense have presented compelling evidence, but after careful deliberation you think

_____

[1] "Public Education Finances," Census Project Update. Available at http://www.census.gov/mp/www/cpu/factoftheday/010196.html.

the chance that the accused is innocent is just 5 percent. (That is, the probability of guilt is .95.) Bearing in mind the common standard that a person must be judged guilty beyond a reasonable doubt, would you vote to convict? Why?

_____

_____

_____

_____

_____

b. Same situation, but now this state has no death penalty, only life in prison without the possibility of parole. Do you convict or not? Why?

_____

_____

_____

_____

_____

c. Change the circumstances. The defendant is accused of a misdemeanor: stealing an iPod from his roommate. The penalty upon conviction is at most six months in prison and a $500 fine. Given the evidence in the case, you assess the probability of guilt as .001. Do you vote for a conviction? Why?

_____

_____

_____

_____

_____

Exercise 12–6. For this exercise you will need appendices A and B in the text. For each of the following questions, use the appropriate appendix to find the answer.

a. Find the probability associated with a z score of 1.20.

_____

b. Find the probability associated with a z score of 2.25.

_____

c. Find the *z* score associated with a probability of .2912.

_____

d. Find the *z* score associated with a probability of .0062.

_____

e. Find a *t* score using a two-tailed test, an alpha level of .05, and 10 degrees of freedom.

_____

f. Find a *t* score using a one-tailed test, an alpha level of .01, and 15 degrees of freedom.

_____

g. Why is the largest probability listed in appendix A .5000?

_____

_____

h. Why do the probabilities in appendix A get smaller as the *z* scores get larger?

_____

_____

i. Why are *t* scores larger with fewer degrees of freedom and smaller with more degrees of freedom?

_____

_____

j. Find the probability associated with a *z* score of 1.96. Find *t* scores using the two-tailed test, a .05 alpha level, and an infinite degree of freedom. What is the important relationship among these answers?

_____

_____

_____

_____

Exercise 12–7. There are many different ways to test hypotheses. In chapter 12 the authors explain how to use a two-sided and a one-sided *t*-test. In the space following, please explain the circumstances under which you would use a one-sided *t*-test and a two-sided *t*-test. Next, use the hypothetical sample data in table 12–1 on the following page to perform a sample *t*-test testing the sample mean against a population mean of 1. Use the 95% confidence level, and calculate the critical *t* value using both a one-sided and two-sided test. How do these calculations illustrate the difference between the two tests?

_____

_____

TABLE 12–1

Demonstration Data for *t*-Tests

i	x
1	4
2	3
3	6
4	0
5	2
6	3

Exercise 12–8. Suppose you have collected sample data on the number of bills written by members of Congress in a single session of Congress. Using the hypothetical sample data in table 12–2, use a *t*-test to determine whether the following null hypothesis is correct: the observed mean in your sample data is not statistically significantly different from the population mean of 5.5. Use the two-tailed test and the 95% confidence level when answering this question. Do you accept or reject the null hypothesis? Why would you use a two-tailed test?

TABLE 12–2

Bills Written by Members of Congress in One Session of Congress

Observation	Bills Written
1	4
2	10
3	1
4	3
5	8
6	7
7	5
8	2
9	3
10	5

Exercise 12–9. Imagine that you have access to hypothetical data from the US State Department about the amount of money different nations spend on educational programs designed to foster goodwill with citizens from other countries. While you would prefer full access to the data, the State Department agrees to tell you only the population mean and provide you with a sample of the population data. You reluctantly agree pending verification that the sample data is representative of the population data. The population mean is $13.4 million. The sample data to which you have access is listed in table 12–3 on the following page. Use a *t*-test to determine whether the sample data to which you have access is representative of the population data by testing the mean of your sample data against the population mean provided by the State Department. You should use a two-tailed test and a 95% confidence level. Explain how a *t*-test could help you decide whether using the sample data is sufficient.

Exercise 12–10. Suppose that another student in the class did not understand how the difference of the means, the standard deviation or the sample size affects hypothesis testing when using a *t*-test or a *z*-test. In the space provided, please explain how each affects the likelihood of finding statistical significance.

a. A larger standard deviation

_____

_____

_____

b. A larger sample size

_____

_____

_____

c. A larger difference of the means

_____

_____

_____

d. What is the standard error of the mean? What does it tell us about our data, and why is it important in testing hypotheses?

_____

_____

_____

_____

_____

_____

TABLE 12–3

Millions of Dollars Spent Fostering Goodwill

Country	Millions of Dollars
Brazil	8
Canada	16
Denmark	11
Ghana	5
Kuwait	6
India	9
Malaysia	1
South Africa	8
South Korea	10
United Kingdom	18

Exercise 12–11. We can use a confidence interval to determine the range in which we expect to find the population mean given a certain level of confidence. The range we find using this statistic varies based on the values of the standard deviation, the sample size, and the confidence level. In the following problems you will manipulate the values used in a population confidence interval to learn how changes in these values affect the size of the confidence interval. For each of the following problems use a two-tailed test.

Change in Standard Deviation

In this step you are going to analyze the effect of changing the size of the standard deviation. Calculate the population confidence interval with a mean of 5, a sample size of 500, a confidence interval of 95%, and a standard deviation of 2. Next, use the same sample size and confidence level but use a standard deviation of 4. How does a larger (smaller) standard deviation affect the calculation of a confidence interval? Why?

_____

_____

_____

_____

Change in Sample Size

In this step you are going to analyze the effect of changing the sample size. Calculate the population confidence interval with a mean of 5, a sample size of 100, a confidence interval of 95%, and a standard deviation of 2. Next, use the same standard deviation and confidence level but use a sample size of 1,000. How does a larger (smaller) sample size affect the calculation of a confidence interval? Why?

_____

_____

_____

_____

Change in Confidence Level

In this step you are going to analyze the effect of changing the confidence level. Calculate the population confidence interval with a sample size of 500, a confidence interval of 90%, and a standard deviation of 2. Next, use the same sample size and standard deviation but use a confidence level of 99%. How does a larger (smaller) confidence level affect the calculation of a confidence interval? Why?

TABLE 12–4

Candidate Cash on Hand

i	x
1	12
2	3
3	6
4	7
5	5
6	10
7	6

_____

_____

_____

_____

_____

Exercise 12–12. In chapter 12, the authors explain how you can use a confidence interval to test a hypothesis. Imagine that you are working on a research project on the effectiveness of Twitter as a political tool. As part of this project, you have hypothesized that the typical politician has nine followers at the beginning of a campaign. Using the hypothetical sample data in table 12–4, where i represents the rows of data and x represents the number of followers each hypothetical candidate has at the beginning of a campaign, calculate a sample confidence interval and decide whether you should accept or reject your hypothesis and explain why you would do so. Use the .05 level and a two-tailed test.

_____

_____

_____

CHAPTER 13

Investigating Relationships between Two Variables

Chapter 13 takes up the analysis of relationships between two variables. Two variables are statistically related when values of the observations for one variable are associated with values of the observations for the other. This chapter gives you a chance to investigate several aspects of two-variable relationships, including strength, direction or shape, and statistical significance.

We mention throughout the textbook that plenty of career opportunities exist for political scientists who have a reasonable understanding of quantitative methods. So if you know a little about cross-tabulations, regression analysis, and statistical inference, you may find jobs waiting for you in campaigns, government, consulting firms, industry, and a host of other areas. But experience tells us that in almost every instance employers look for well-trained social scientists who can also clearly, succinctly, and forcefully explain numerical procedures and results to people who do not have much knowledge of (and, frankly, not much interest in) these topics. This ability becomes particularly valuable when someone asks, "Is this an important finding, one I should pay attention to, or can I ignore it?" Hence, we encourage you to think carefully about the answers to the questions posed here and in other chapters.

Here we offer some advice for translating statistical results into common sense:

- **Before responding to the question (or submitting a report to your boss), consider what is being asked**. In the real world, people want answers expressed in real-world terms. You have been told that a regression coefficient, for example, can be interpreted as the slope of a line or as indicating how much Y changes for each unit increase or decrease in X. An r measures goodness of fit. But for many people, explanations expressed in those terms might as well be gibberish. It is essential, then, that you make sense of statistical terms by placing them in a specific substantive context. Regression and correlation coefficients indicate how and how strongly one thing is related to another. That's the importance of such coefficients. So make sure that you talk about variables, not equations or Greek letters or abstract symbols. For example, write "Income is related to attitudes on taxation," not "X is related to Y."
- **Go even further to explain the nature of interconnections**. Yes, income is correlated with opinions. But how? You might add "The wealthier people are, the more they favor cutting taxes, but even the lower and middle classes want some degree of tax relief." This statement is much more informative to a nonstatistician than the statement "There's a positive correlation."
- **We have suggested numerous times that the definitions of variables determine how we interpret the phenomena under consideration.** People want to know if there is a meaningful difference between A and B or they to want to understand how strongly X and Y are connected. Numbers alone won't do the job. Only the names and meanings of the variables will. So don't write something like "There is a difference of 10 (or a difference of 10 units)." The difference is in what units? Dollars? Years? Pounds? Percentages?

- **In this same vein, the measurement scales of variables are critically important.** They should be one of the first things you look for and understand. In some cases the meaning of the categories or the intervals will be obvious or intuitive, as in "years of formal schooling." In other instances, a scale may present some more or less subtle explanatory difficulties. Take income, a variable we discuss several times in the text. In many cases the scale is just "dollars," so "$1,000" has a clear meaning. In other instances a variable may be measured in thousands, millions, or even billions of dollars. (That is, "$10" may stand for $10 million.) It is important that you know exactly which scale is being used. After all, a change of 10 means one thing when we are talking about simple dollars and quite another when the scale is millions or billions. And complicating matters even further, social scientists often measure variables on abstract or artificial scales. ("Where would you place yourself on this ten-point thermometer of feelings about the president?") If one person's score is 7 and another's is 5, the scores differ. But how important in the world of politics is this difference? As we emphasize following, it is only possible to give a reasoned judgment; there will seldom be a clearly right or wrong answer.

- **You can help yourself by keeping track of measures of central tendency and variability.** If most respondents in a study have values near the mean or median, then one person whose score is 2 standard deviations away may be unusual and warrant further investigation. Was the individual's score measured correctly? Is he or she an "extremist"? The two interpretations, which have vastly different substantive implications, can be adjudicated only with thought and perhaps further research.

- **Don't let one or two measures (for example, chi-square, r) do all your interpretative work.** Instead, try to examine the data as a whole. If you have a contingency table, look for patterns of association within the table's body. Compare different categories of the response patterns. For example, individuals at the high and low ends of a scale often differ greatly in their attitudes. Those in the middle may be more homogeneous. Or response patterns may differ as you move from one end of the table to the other. Whatever the case, it is important to examine data, such as that found in a table, from several angles. Similarly, variables having quantitative (ratio and interval) scales should be plotted as they are in the text. (Graphing software is so widely available that this shouldn't be a chore.) From such graphics you can determine the form of relationships and their strength and locate "outlying" observations, among other things.

- **As we noted, assessing importance is one of the hardest tasks facing data analysts.** This problem has both statistical and substantive aspects, and both have to be considered simultaneously. Terms such as *statistical significance* and *explained variation* pertain to observed data, not to people's feelings and behavior. Therefore, finding that a chi-square is statistically significant may or may not be important. By the same token, the fact that income "explains 60 percent of the variation in political ideology" doesn't necessarily mean we know much about why people are liberals, moderates, or conservatives. Data analysis helps us understand, but it does not replace hard thought about the substance of a topic.

Here is an example that may tie these ideas together. An investigator wants to know if Americans are more knowledgeable about government and politics than Germans. She conducts a survey of 5,000 citizens in each country (total $N = 10,000$) and discovers that 20 percent of Americans and 25 percent of Germans can name their representatives to the local legislature. Statistically speaking, this would be a highly significant difference. But does it have practical importance? Most observers would probably say, "No, there's no functional or meaningful difference. The statistical significance is a product of the huge sample size."

Exercise 13–1. Fill in the blanks.

a. The relationship between two categorical variables (nominal and/or ordinal) is usually studied by the analysis of _____ tables.

b. Comparison of means and the analysis of variance are used to look for a relationship between two variables when the independent variable is a _____ -level or _____ -level measure and the dependent variable is a _____ -level or _____ -level measure.

c. _____ analysis is used to see if there is a _____ relationship between two variables if both variables are interval- or ratio-level measures.

d. _____ analysis is used to examine the existence of a relationship when both variables are nominal- or ordinal-level measures.

e. In a contingency table, the conventional practice is to make the independent variable the _____ (row/column) variable.

f. If the independent variable is the row variable in a contingency table, then the percentages in each _____ (row/column) should add up to 100.

g. For an investigation of the socioeconomic effects of regime type in developing countries, an investigator has measured the percentage of the population with access to safe drinking water in twelve authoritarian and nine democratic countries. A _____ test would be an appropriate procedure in this situation.

h. A consulting firm wants to know if there is a relationship between religious identification and support for federal funding of stem cell research. It surveys 500 residents in a large metropolitan area and asks respondents to which church they belong (if any) and where they place themselves on a five-point "agree-disagree" scale of support for publicly funded stem cell research. The analysis of these two variables would most likely involve the analysis of _____ .

Exercise 13–2. Statisticians frequently use graphs to visualize relationships. Look at figure 13–1. It displays what we have labeled a strong "negative linear correlation" between Y and X.

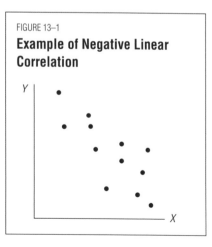

FIGURE 13–1

Example of Negative Linear Correlation

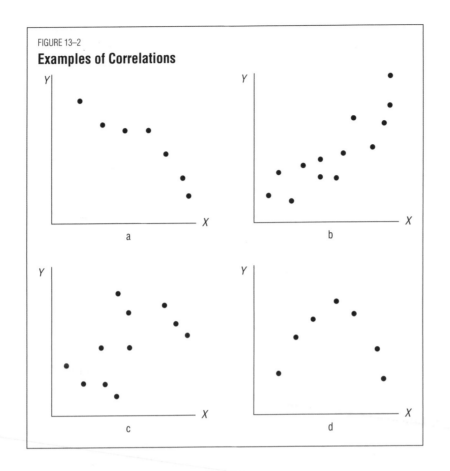

FIGURE 13–2

Examples of Correlations

Describe the type and (approximate) strength of the four relationships shown in figure 13–2.

1. _____

_____

2. _____

_____

3. _____

_____

4. _____

_____

Exercise 13–3. Using the hypothetical data given in table 13–1, create a contingency table to examine the relationship between ethnicity and political participation as measured by voting. Make sure that you include the appropriate percentages as well as the frequencies.

TABLE 13–1

Data on Ethnicity and Voting

Ethnicity	Turnout	*N*
White	Voted	150
White	Did not vote	50
Nonwhite	Voted	30
Nonwhite	Did not vote	20

a. Place your table here.

b. In your opinion, does a relationship exist between voting and ethnicity? Briefly explain.

_____

_____

_____

_____

_____

c. Thinking of percentages in probability terms, what is the *probability* of a nonwhite person voting? _____ Of a white person? _____

d. What are the *odds* of a white person voting as opposed to not voting? _____ The odds that a nonwhite person will vote as opposed to not vote? _____

e. What is the odds ratio? _____ Briefly interpret this number.

_____

_____

_____

_____

_____

Exercise 13–4. Table 13–2 is a contingency table from a computer program that shows the relationship between region and the political parties of ninety-nine US senators. (One independent senator is not included.)

TABLE 13–2

Cross-Tabulation of Senators' Party Identification and Region

Party Identification	Region				
	1	2	3	4	Total
1 Democrat	16	9	14	10	49
% within region	69.6%	37.5%	53.8%	38.5%	49.5%
2 Republican	7	15	12	16	50
% within region	30.4%	62.5%	46.2%	61.5%	50.5%
Total	23	24	26	26	99
% within region	100.0%	100.0%	100.0%	100.0%	100%

Source: Computer-generated table.

a. Do the data indicate that Democratic and Republican senators tend to come from different regions of the country?

_____

_____

b. If you did not know which region a senator was from, what would be your best guess of his or her party affiliation?

_____

_____

c. What would be an appropriate measure of association? Why?

_____

_____

_____

d. Calculate the measure.

_____

_____

_____

Exercise 13–5. Examine the data presented in table 13–3.

TABLE 13–3

Cross-Tabulation of Prayer by Political Ideology

PRAY: How often does R pray	Political Orientation			Total
	1 Liberal	2 Moderate	3 Conservative	
1 Several times a day	33	71	91	195
	18.0%	23.9%	41.7%	27.9%
2 Once a day	53	86	64	203
	29.0%	29.0%	29.4%	29.1%
3 Several times a week	22	48	25	95
	12.0%	16.2%	11.5%	13.6%
4 Once a week	17	18	14	49
	9.3%	6.1%	6.4%	7.0%
5 Less than once a week	55	72	23	150
	30.1%	24.2%	10.6%	21.5%
6 Never	3	2	1	6
	1.6%	.7%	.5%	.9%
Total	183	297	218	698
	100.0%	100.0%	100.0%	100.0%

Source: Computer-generated table.

a.   Which is the independent variable? Which is the dependent variable? Describe the relationship between the variables.

b.   The gamma value for the data in the table is −.288. What does gamma tell you about the relationship between political orientation and frequency of praying? Why does gamma have a negative sign?

Exercise 13–6. The analysis of data with a chi-square test is widely used in public opinion and other types of empirical research. But the results are not self-explanatory; it takes experience to get a feel for what the procedure indicates. Fortunately, the Internet provides access to all sorts of teaching tools that can help you grasp the meaning of statistics such as chi-square. As an example, a nifty applet (a small software program that you download from the Internet to your desktop) available at http://home.ubalt.edu/ntsbarsh/business-stat/other applets/catego.htm allows you to enter raw frequencies from a contingency table (with up to six rows and six columns) to find chi-square values. Just fill in the table—*remember to enter raw frequencies, not percentages or relative frequencies*—and click on "Calculate." The applet reports the chi-square statistic and the attained probability (*p*-value) under the null hypothesis of statistical independence. Figure 13–3 shows how the screen appears before data have been entered and after we supplied a few hypothetical frequencies. Ignore the correlation coefficient, and of course interpret the results in *your* words.

FIGURE 13–3

Chi-Square Applet Screenshot

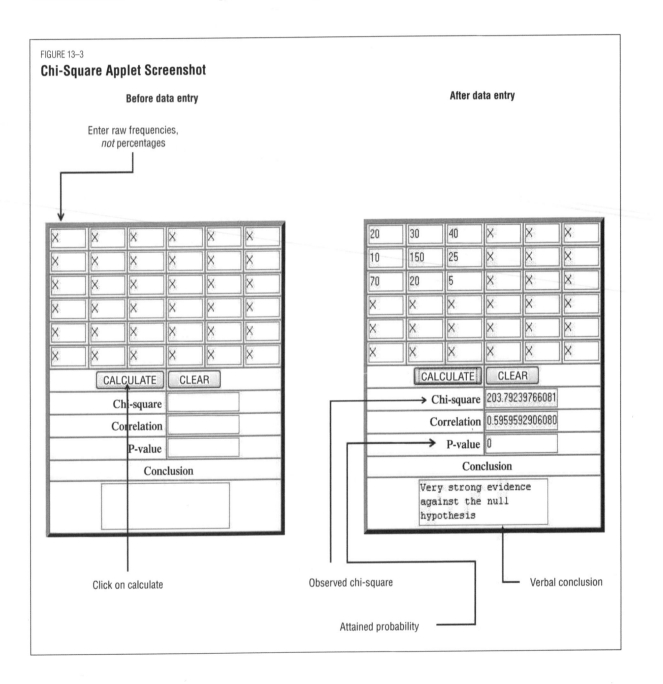

a. Use the applet described in this exercise to enter the values from the following table:

	X		
	50	50	50
Y	50	50	50
	50	50	50

What is the chi-square? Is it statistically significant? If so, at what level?

_____

_____

b. Enter the values from the following table (make sure zero frequencies are typed in when appropriate):

	X		
	50	0	0
Y	0	50	0
	0	0	50

What is the chi-square? Is it statistically significant? If so, at what level?

_____

_____

c. Enter the values from the following table (make sure zero frequencies are typed when appropriate):

	X		
	50	40	40
Y	40	40	40
	40	40	40

What is the chi-square? Is it statistically significant? If so, at what level? Is there a relationship?

_____

_____

d. Enter the values from the following table (make sure zero frequencies are typed in when appropriate):

	X		
	500	400	400
Y	400	400	400
	400	400	400

What is the chi-square? Is it statistically significant? If so, at what level?

_____

_____

e. What feature or property of chi-square do the previous two tables illustrate?

_____

_____

_____

_____

Exercise 13–7. A major and continuing controversy in American politics has been the privatizing of governmental functions. Some states, for example, have turned to private contractors to house and "rehabilitate" their prisoners. Advocates claim that this practice saves the taxpayers money. Critics, however, say that these institutions, which are run on a for-profit basis, cut corners in inmate care and supervision.

TABLE 13–4

Inmate Deaths by Facility

Type of Facility	Mean	Number of Institutions
Federal	8.86	(112)
State	19.48	(1,225)
Private	4.65	(110)
Total	17.53	(1,447)

Source	Sum of Squares	Degrees of Freedom	Mean Square	Observed F	Probability
Explained (Type of prison)	31,341.430	2	15,670.715	7.490	.0006
Error	3,020,990.558	1,444	2,092.099		
Total	3,052,331.988	1,446			

Your employer, a nonprofit criminal justice organization, has funded a study of the issue. A small part of the resulting data appears in the next two tables. Your boss asks you to make sense of these numbers. This portion of the analysis involves two variables collected by the Bureau of Justice Statistics: (1) the operator of the facility, a federal or state agency or a private corporation, and (2) answers to the following question asked of supervisors: "Between July 1, 1994, and June 30, 1995, how many total inmates died while under the jurisdiction of this facility?"

a. What type of analysis is this? Regression, analysis of variance, or what?

_____

b. What is the objective of this analysis?

_____

_____

c. Did type of the facility (federal, state, private) have an effect on death rates among inmates? Explain.

_____

_____

d. Look at the data in both tables. Can you translate the problem into a statistical hypothesis? Explain it to your employer.

_____

_____

e. Assuming these data came from a random sample of prisons in the United States (they did not), could the differences between the death rates be attributable to sampling error or is there evidence that the type of operator has an effect on mortality? Explain. (*Hint:* This is a question of statistics, not substance.)

_____

_____

f. Interpret these data and statistical results substantively, that is, in a way that makes sense to journalists and public officials. Try to think of a reason why the statistical results reported in item 13–7e came out as they did. Are we observing a direct connection between type of prison and inmate death rates, or are there other variables that ought to be taken into account? Does the study throw any light on the policy debate about privatization?

_____

_____

_____

_____

Exercise 13–8.   We hear a lot of discussion about "group ratings" of politicians and candidates. The American Civil Liberties Union (ACLU), for example, annually rates members of Congress on their support and opposition to legislation about civil liberties (for example, freedom of speech). Those senators and representatives who vote against laws and amendments banning flag burning get higher scores than those who do not.

Following the instructions provided to you by your instructor regarding which statistics software package to use, access the file "Senate.dat" on the Web site for this book (http://psrm/cqpress.com) and examine the relationship between a senator's party affiliation and score on the ACLU scorecard. Use the variable "PARTYRD," which excludes independents from the analysis, as the measure of senators' party affiliations.

a. What level of measurement is PARTYRD?

_____

b. What is the level of measurement for ACLU?

_____

c. Write a hypothesis about the relationship between party and ACLU scorecard scores.

_____

_____

d. What statistical procedures would you use to see whether a relationship exists between the variables and how strong a relationship it is?

_____

_____

_____

_____

_____

e. It is *not* appropriate to use a test of statistical significance for the data. Why not?

_____

_____

f. Run the appropriate statistical analyses and then discuss your results in substantive terms.

_____

_____

_____

_____

_____

_____

_____

_____

Exercise 13–9.   Do political party leaders represent their rank-and-file members? Do their opinions and stands on issues generally agree with those of their supporters? Or are they more liberal or conservative? Richard Herrera investigated this question by comparing the mean views of delegates to the 1988 Democratic Party convention, presumably a good cross-section of party leaders, with those of Democratic voters.[1] He hypothesized that if leaders are out of touch with followers, there will be a difference between their average issue positions. A small portion of his results can be seen in table 13–5.

---

[1] Richard Herrera, "Are 'Superdelegates' Super?" *Political Behavior* 16 (March 1994): 79–92.

TABLE 13–5

Mean Views of Democratic Delegates and Democratic Voters, 1988

Issue	Mean Delegate View	Mean Partisan Voter View	Mean Difference and Significance
Defense spending	2.33	3.11	−.78**
Get along with Russia	2.24	3.23	−.99**
Government helps blacks	2.86	1.92	−.94**
Place of women	1.47	2.46	−.99**

Source: Richard Herrera, "Are 'Superdelegates' Super?" *Political Behavior* 16 (March 1994): 88.

** = Significant at .01 level.

The mean delegate and partisan voter responses are given on seven-point scales. For instance, both delegates and a random sample of Democratic voters were asked about their opinions on defense spending. A 1 on the scale represents the most liberal position (that is, cut defense spending), whereas 7 is the most conservative stance. If everyone in a particular group were liberal, the mean score would be 1. The other questions, for which respondents could reply on similar seven-point scales, were as follows: "Should we try to get along with Russia?" (1), or is this a "big mistake"? (7); "Should the government in Washington make every effort to improve the social and economic position of blacks?" (1), or should there be "no special effort"? (7); and "Should women have an equal role with men in running business, industry, or government?" (1), or is women's "place in the home"? (7). The last column represents the difference in means between delegates and voters. The author writes that the symbol ** means "difference is significant at p .01."[2] Try to make sense of these results.

a. What general hypothesis is the author investigating?

_____

_____

_____

b. What statistical hypothesis does each line of the table test?

_____

_____

_____

---

[2] Although it may be losing favor, the use of stars or other typographical symbols has, in the past, been a common way to report significance. An asterisk (*), for instance, conveys the idea that the statistical hypothesis (for example, statistical independence) has been rejected at the .05 level. But it is far preferable for you to indicate the attained or observed probability of the sample result if the null hypothesis is true. It's analogous to a friend's telling you that the Baltimore Orioles beat the New York Yankees. That may be good or bad news, but the next question is, "Okay, what was the final score?" So rather than reporting that a result is significant at the .05 level, indicate as closely as you can the actual probability under the null hypothesis. It might be, say, .04, which is barely significant at the .05 level, or it might be .002, which is highly significant at that level. Most software reports the attained probability, so why not include it in your report?

c. Which of these statistical hypotheses should Herrera reject and why?

_____

_____

_____

d. So far you have considered statistical hypotheses: Should they be rejected or not? Now translate these results into terms an average citizen can understand. What, in short, do the results say about Democratic leaders and average party voters? (*Hint:* Always keep in mind the meaning of the questions and the scales. In other words, if one person has a higher score than another individual on, say, the defense spending issue, would you say the first was more liberal or conservative than the second?)

_____

_____

_____

_____

_____

Exercise 13–10.   Referring to the research of several political scientists, David Brooks, a *New York Times* columnist, writes, "Party affiliation even shapes people's perceptions of reality. . . . People's perceptions are blatantly biased by partisanship."[3] The 2000 American National Election Study located on the Web site that accompanies this workbook (http://psrm/cqpress.com) provides data to verify this assertion. To do this assignment, you will need software that can cross-tabulate two variables and calculate elementary statistics. Consult your instructor or laboratory assistant for help. The Web site contains a guide to this data set ("anes-2000readme.txt"). If you refer to it, you will find that "Party Identification" (variable 24) measures both the strength and direction of political party affiliation. Other variables represent measures of perceptions. Try to see if Brooks's statement holds water.

a. At the time of the 2000 presidential election, a Democrat, Bill Clinton, occupied the White House. Voters rightly or wrongly connect the past performance of the economy to the presidency. So use the variable "Economy During Past Year" (variable 19) to form a contingency table or cross-tabulation of party identification and perceptions of economic performance. Print or place a properly labeled and formatted table in the space following. (*Hint:* You don't have to enter the raw frequencies in each cell, just the appropriate percentages. But you should include the marginal frequencies for the independent variable.)

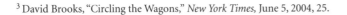

[3] David Brooks, "Circling the Wagons," *New York Times,* June 5, 2004, 25.

b. What are the chi-square and degrees of freedom for this table? What does the chi-square mean in statistical *and* substantive terms?

_____

_____

_____

_____

_____

c. What is gamma? _____ What is Kendall's tau-*b*? _____ What do these measures tell you?

_____

_____

_____

d. Now use the variable "Gore: Leader?" (variable 27) to test the hypothesis about the effects of partisanship on perceptions. In the following space, print or carefully place a table that shows the relationship between party identification and beliefs about Al Gore being a "strong leader."

e. What are the chi-square and degrees of freedom for this table? What does the chi-square mean in statistical terms? What is the practical or theoretical import of the findings? As an example, what do the results imply about the effectiveness of television commercials that attack a candidate's leadership abilities? Are they likely to "work" on all voters equally? Under what conditions?

_____

_____

_____

_____

f. What is gamma? _____ What is Kendall's tau-*b*? _____ What do these measures tell you?

_____

_____

_____

_____

Exercise 13–11.   It might be instructive to explore the issue in exercise 12–12 in a comparative context. The data from the "British Election Study 2001" offer a chance to do so. These data are available on the Web site for this book (http://psrm/cqpress.com) in a couple of formats: "bes2001.dat" and "bes2001.por," an SPSS file. A text file, "bescodebook.txt," describes the variables.

At the time of the 2001 general elections, Tony Blair's Labour Party controlled Parliament and hence the government. There are two other major parties in Britain, the Conservative and the Liberal Democrat Parties. Try to determine if party identification (variable 1) is related to variable 6, "How do you think the general economic situation in this country has changed over the last 12 months?"

(*Hint:* As in the United States, Britain has several "minor" or "lesser" parties. To simplify your analysis we suggest you eliminate all the party identifiers other than Labour, Conservative, and Liberal Democrat.)

a. In the space below print or place the contingency table or cross-tabulation.

b. What is your interpretation? Do the data support the assertion that party identification colors perceptions of political and economic affairs?

_____

_____

_____

_____

_____

_____

c. What are the chi-square and degrees of freedom for this table? What do they mean in statistical and substantive terms?

_____

_____

_____

_____

_____

_____

d. Obtain or calculate lambda and interpret (both statistically and substantively) its meaning in this context.

_____

_____

_____

_____

_____

_____

Exercise 13–12.   At the heart of a regression analysis is the concept of minimizing the squared errors. Using figure 13–13 in the textbook and the discussion about it, answer the following questions.

a. Explain in plain English how a regression minimizes the squared errors. Why does a regression give us a single line that best fits the data?

_____

_____

_____

_____

b. What is the difference between an observed value of Y and a predicted value of Y?

_____

_____

_____

_____

c. Why would one expect to see all of the data points on the line representing the mean and the regression line if an independent variable perfectly predicts a dependent variable?

_____

_____

_____

_____

Exercise 13–13. The following hypothetical data can be used to analyze political communication among friends. In this example, we try to assess how Facebook friends communicate political information. We have data on the number of times each Facebook user was exposed to political information from a friend and how many times that user shared political information with a friend. We suspect that an increase in exposure to political information causes an increase in the sharing of political information.

Use the data in table 13–6, where "Exposure to Political Info" is the independent variable and "Political Info Shared with Another" is the dependent variable, to plot the regression line. Remember that in order to plot a regression line you will first need to perform all of the underlying calculations to find b, a, and predicted values of Y.

TABLE 13–6

Exposure to and Sharing of Political Information

i	Exposure to Political Info	Political Info Shared with Another
1	2	2
2	1	3
3	1	1
4	5	8
5	4	3
6	7	5
7	7	7
8	2	1
9	3	6
10	0	0

Exercise 13–14. Define each of the following and explain how you would use each in an analysis of a bivariate relationship.

■ R^2
■ r

- Y
- Y_i
- α
- β
- ε

Exercise 13–15. Expenditures for education are a continuing controversy in the United States. Many people think more needs to be spent to improve the quality of education, while others feel that just "throwing money" at schools is no solution at all. Needless to say, we cannot delve deeply into this issue here, but table 13–7, supplied by Professor Deborah Lynn Guber of the University of Vermont, presents some relevant data.[4] In particular, it lists for each state:

- Current expenditure per pupil in average daily attendance in public elementary and secondary schools, 1994–1995 (in thousands of dollars)
- Average verbal SAT score, 1994–1995
- Percentage of all eligible students taking the SAT, 1994–1995

For this exercise we examine only the relationship between verbal SAT scores and per-pupil expenditures. For this analysis, we suggest the use of a computer program.

USING ONLINE SOFTWARE

If you can choose the software and have no strong preference, we suggest that you look at one of the dozens and dozens of online statistical programs that are widely available. An excellent system, for example, is "VassarStats" (http://faculty.vassar.edu/lowry/Vassar Stats.html), prepared and maintained by Professor Richard Lowry of Vassar College.

To use Professor Lowry's regression program, go to the site, click on "Correlation & Regression" in the left frame, scroll down to "Basic Linear Correlation and Regression," and then, depending on your preference, click on "Data-Import Version" or "Direct-Entry Version." If you choose the latter, enter the number of cases or observations in the next window, and then type in the data in the box provided in the following window, making sure that Xs are entered in the first column and Ys in the second. When finished, click on "Calculate."

Alternatively, you can use the data import option. This allows you to copy adjacent columns of data from a spreadsheet or word processor. First, use a word processor or spreadsheet program to enter your data in columns. Place X values in the first column and Y values in the next. Then copy them (in Word, you hold down the "Alt" key to copy blocks of text) and paste them in the "Data Entry" window. Click on "Calculate." In both cases, the application provides all the information needed for the assignment except a scatterplot.

---

[4] The data appear on the archive Web site of the *Journals of Statistics Education* (www.amstat.org/publications/jse/jsedataarchive.html). They and supporting material were submitted by Deborah Lynn Guber, University of Vermont.

TABLE 13–7

Per-Pupil Expenditures, Verbal SAT Scores, and Percentage of Students Taking the SAT, by State

State	Per-Pupil Expenditures (in thousands of dollars)	Verbal SAT Score	Percentage of Students Taking the SAT
Alabama	4.405	491	8
Alaska	8.963	445	47
Arizona	4.778	448	27
Arkansas	4.459	482	6
California	4.992	417	45
Colorado	5.443	462	29
Connecticut	8.817	431	81
Delaware	7.030	429	68
Florida	5.718	420	48
Georgia	5.193	406	65
Hawaii	6.078	407	57
Idaho	4.210	468	15
Illinois	6.136	488	13
Indiana	5.826	415	58
Iowa	5.483	516	5
Kansas	5.817	503	9
Kentucky	5.217	477	11
Louisiana	4.761	486	9
Maine	6.428	427	68
Maryland	7.245	430	64
Massachusetts	7.287	430	80
Michigan	6.994	484	11
Minnesota	6.000	506	9
Mississippi	4.080	496	4
Missouri	5.383	495	9
Montana	5.692	473	21
Nebraska	5.935	494	9
Nevada	5.160	434	30
New Hampshire	5.859	444	70
New Jersey	9.774	420	70
New Mexico	4.586	485	11
New York	9.623	419	74
North Carolina	5.077	411	60
North Dakota	4.775	515	5
Ohio	6.162	460	23
Oklahoma	4.845	491	9
Oregon	6.436	448	51
Pennsylvania	7.109	419	70
Rhode Island	7.469	425	70
South Carolina	4.797	401	58
South Dakota	4.775	505	5
Tennessee	4.388	497	12
Texas	5.222	419	47
Utah	3.656	513	4
Vermont	6.750	429	68
Virginia	5.327	428	65
Washington	5.906	443	48
West Virginia	6.107	448	17
Wisconsin	6.930	501	9
Wyoming	6.160	476	10

a. Prepare a scatterplot of verbal SAT scores (Y) against per-pupil expenditures. Attach it to the assignment. Can you discern a relationship?

_____

_____

b. What is the correlation coefficient? What does it tell you about expenditures and SAT scores? Is there anything surprising or interesting in this result? Explain.

_____

_____

_____

_____

_____

c. What is the estimated regression equation?

_____

_____

d. Interpret the regression coefficient statistically and substantively.

_____

_____

e. Is the regression coefficient statistically significant? At what level? Explain.

_____

_____

Exercise 13–16. Suppose a political scientist hypothesizes that lower economic growth rates will be associated with greater political instability. His thinking rests on this logic: in a world of satellite television, the Internet, and video cell phones, people increasingly expect and demand the living standards they see in developed nations. If national productivity does not bring these rewards, citizens become frustrated and place heavy demands on the political system. Instability, even violence may result. If, on the other hand, the economy can meet demands for higher living standards, government will have greater legitimacy. To test this idea the researcher collects data for a sample of 34 nations. The dependent variable, a measure of political stability, is a percentile that ranges from 0 to 100, with 0 meaning the least stability and 100 the most. The independent variable is the percentage per capita growth rate in gross domestic product (GDP) in 2005.[5] The investigator, who is writing an article for publication, claims that these data support his proposition. But he asks for your opinion before sending the paper to a journal. Answer the questions following and render an opinion.

[5] GDP data are from the *CIA World Fact Book.* Available at https://www.cia.gov/cia/publications/factbook/fields/ 2003.html.

Table 13–8 presents the results of a regression analysis of political stability on GDP growth in a format that commonly appears in scholarly journals. Not as much information seems to be supplied. But in fact you can reconstruct the significance tests. Assume that the null hypotheses pertaining to the regression parameters are that both βs equal zero. (As we say in the text, these are the usual null hypotheses.) Note that the standard errors appear in parentheses below the estimated coefficients. Hence, you can compute the observed t values. Similarly, the degrees of freedom and the critical t (for any level of significance listed in appendix B in the text) can be determined because you have the sample size.

TABLE 13–8
Regression Analysis of Political Stability on GDP Growth

	Estimated Constant and Coefficient (standard error)
Constant	79.31**
	(12.36)
Growth in GDP	−6.569*
	(2.707)

Source: CIA World Fact Book, 2007.
$N = 34$, $R^2 = .155$.
*Significant at .05. **Significant at .001.

a. Compute the observed t for the regression constant.

_____

_____

b. Compute the observed t for the regression coefficient.

_____

_____

c. What are the degrees of freedom for testing the significance of the regression coefficient?

_____

_____

d. What is the critical t at the .05 level for a two-tailed test?

_____

_____

e. The asterisk in the table suggests that $\hat{\beta} = -6.569$ is statistically significant. Is this assertion correct? Why or why not?

_____

_____

f. What proportion of the variation in political stability is "explained" by growth in GDP?

_____

_____

g. What is the correlation coefficient (r) between stability and growth? (*Hint:* What is the connection between R^2 and r?)

_____

_____

h. In sum, do the data support the political scientist's research hypothesis (not the null or alternative hypothesis) about the effect of economic change on political stability? Explain your answer. (*Hint:* Look carefully at the regression coefficient and think about what the number means in substantive or practical terms.)

_____

_____

_____

_____

Exercise 13–17.   For this exercise you will test the hypothesis that older citizens hold the Supreme Court in higher regard than do younger citizens, using a difference of means test. The data include two random samples drawn independently, and you should assume the variances are equal. Sample 1 includes 30 citizens thirty years of age or older with a sample mean of 87 and a population standard deviation of 4. Sample 2 includes 25 citizens under age thirty with a sample mean of 85 and a population standard deviation of 3. Choose a confidence level with which you are comfortable and explain why you chose that level, then decide if you accept the hypothesis. (*Hint:* Remember that the variance is the square of the standard deviation.)

_____

_____

_____

_____

_____

CHAPTER 14

Multivariate Analysis

In chapter 14 we present several different statistical methods, or models, for investigating relationships that involve a dependent variable and more than a single independent variable. Multivariate analysis helps researchers to discover spurious relationships, to measure the effect of changes in multiple independent variables on a dependent variable, and, generally, to strengthen claims about causal relationships. Multivariate statistical procedures allow researchers to control statistically for other factors instead of controlling other factors experimentally. Substitution of statistical control for experimental control is not a perfect solution in the quest to establish causal explanations for political phenomena, and, therefore, results and claims based on statistical evidence must be scrutinized carefully.

As with bivariate data analysis, the choice of analytical procedure depends on the way variables, particularly the dependent variable, are measured:

- Contingency table analysis is used when the data are categorical, that is, when the variables have categories. This method quickly becomes unwieldy and difficult to interpret as the number of tables and cells increases and the number of cases or observations in cells decreases.
- Linear multiple regression investigates whether there is a linear relationship between a numerically measured dependent variable and multiple independent variables and allows researchers to assess how much a one-unit change in an independent variable changes the dependent variable when all the other variables have been taken into account or controlled. So-called dummy variables are used when there are categorical independent variables.
- A logistic regression model is used when the dependent variable is dichotomous (that is, a binary variable with values of 0 or 1) and estimates the probability that the dependent variable equals 1 as a linear function of the independent variables.

Also, as with bivariate data analysis, we are interested in the strength of the relationship, or how well the model fits the data, and in the statistical significance of our findings.

Although some models and statistical procedures can get quite complicated, with practice you will be able to use at least the simpler ones and to interpret their results. The exercises give you practice in deciding which procedure is appropriate to use, how to relate your data to the procedure, and how to interpret the results of the analyses.

Exercise 14–1. Here are a few questions to spark your curiosity. Following are descriptions of possible causal relationships. For each one identify the causal assertion and draw an arrow diagram similar to the ones used in chapters 13 and 14 to represent it. Then, think of a third variable (or variables) that might explain the original relationship. Draw another arrow diagram to illustrate the possible three-variable association. Here is an example. Suppose someone tells you, "There is a statistically significant relationship between foot size and the number of words in the vocabularies of the students in the Denver, Colorado, public school system. Children with big feet tend to know many more words than children with dainty feet."[1] The statement asserts a relationship between foot size and size of vocabulary. If the relationship were causal, it could be diagrammed as shown in figure 14–1.

FIGURE 14–1

X
(foot size) ——————————————→ Y
(number of words
in vocabulary)

It is, of course, not clear that the size of anyone's feet causes vocabulary acquisition. Instead, the relationship probably reflects aging: as children get older, their feet grow larger *and* they develop larger vocabularies (figure 14–2).

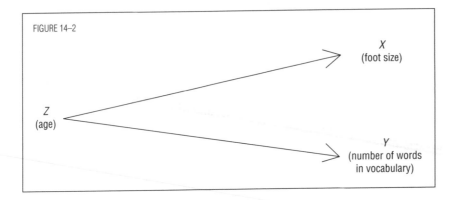

FIGURE 14–2

Z
(age)

X
(foot size)

Y
(number of words
in vocabulary)

The key to answering these questions is to think and reason carefully.

a. "Should violence on cable television be more closely regulated by government? The *Washington Post* reports that the Federal Communications Commission found that 'a *correlation* exists between bloodshed on television and violence in real life. . . .'"[2] Think of a factor that might explain the relationship and draw the two diagrams. Briefly discuss your ideas.

---

[1] Stimulated by an example at "Spurious Correlation and Its Detection," *Autobox.com*. Available at http://www.autobox .com/spur10 .html.
[2] John Dunbar, "FCC: Govt. Could Regulate TV Violence," *Washington Post*, April 25, 2007. Available at http://www.washingtonpost.com/ wp-dyn/content/article/2007/04/25/AR2007042502332.html.

_____

_____

_____

_____

b. *Claim:* There is a strong and statistically significant positive correlation between the percentage of Hispanics in a city and the percentage of students who do not complete high school. What is the causal assertion, and what is a possible alternative?

_____

_____

_____

_____

c. "The [Dartmouth Medical School] team found a direct correlation between the number of smoking scenes watched and the chances of becoming a habitual smoker: Children who had seen the most scenes were twice as likely to end up addicted as those who had seen the fewest. . . . The study is in the September issue of the *Archives of Pediatrics and Adolescent Medicine.*"[3] What is the causal claim? Can you think of an alternative explanation? Illustrate your discussion with diagrams.

_____

_____

_____

_____

d. Senator Sam Brownback, Republican, said in a televised debate on September 5, 2007, "In countries that have redefined marriage, where they've said, 'Okay, it's not just a man and a woman, it can be two men, two women,' the marriage rates in those countries have plummeted to where you have countries now in

_____

[3] Rick Weiss, "Science Notebook," *WashingtonPost.com,* September 24, 2007. Available at http://www.washingtonpost .com/wp-dyn/content/article/2007/09/23/AR2007092300770.html.

northern Europe where 80 percent of the firstborn children are born out of wedlock. . . . And currently in this country—currently—we're at 36 percent of our children born out of wedlock."[4] What claim is Senator Brownback making? There may be several, but just pick one—try to recast it as a causal assertion and think of a possible alternative explanation for the relationship he asserts.

e. An article on the Web site of the Heritage Foundation, a Washington, D.C., policy organization ("think tank"), asserts, "Many lawmakers in Congress and in the states assume that the high level of crime in America must have its roots in material conditions, such as poor employment opportunities and a shortage of adequately funded social programs. . . . The central proposition in official Washington's thinking about crime is that poverty is the primary cause of crime." The author of the article disagrees with this statement and argues instead that the number one cause of the development of criminal behavior is fatherless families. He writes, "Over the past thirty years, the rise in violent crime parallels the rise in families abandoned by fathers. . . . High-crime neighborhoods are characterized by high concentrations of families abandoned by fathers. . . . State-by-state analysis by Heritage scholars indicates that a 10 percent increase in the percentage of children living in single-parent homes leads typically to a 17 percent increase in juvenile crime. . . . The rate of violent teenage crime corresponds with the number of families abandoned by fathers."[5] Here you have an implicit debate between "many lawmakers" and researchers at the Heritage Foundation and elsewhere. Use arrow diagrams to illustrate their causal arguments. Then try to think of a way of reconciling these positions. Is it possible, in other words, that both sides are partly correct? Again use diagrams to help clarify your thinking.

Exercise 14–2.   Here are a few research hypotheses and designs. For each one, explain what would be an appropriate tool for statistical analysis and why. (*Hint:* What are the dependent and independent variables and how are they measured?)

a. An investigator believes acts of terrorism are caused mainly by sudden decreases in the economic standard of living of large numbers of society members.[6] Data collected from fifty-five nations consist of the

---

[4] "The Claim," *WashingtonPost.com*, September 19, 2007. Available at http://www.washingtonpost.com/wp-dyn/content/ article/2007/09/18/AR2007091801864.html.

[5] Patrick F. Fagan, "The Real Root Causes of Violent Crime: The Breakdown of Marriage, Family, and Community," *The Heritage Foundation*, March 17, 1995. Available at http://www.heritage.org/Research/Crime/BG1026.cfm.

[6] Incidentally, for a classic statement of this hypothesis see Crane Brinton, *The Anatomy of Revolution*, rev. and exp. ed. (Englewood Cliffs, N.J.: Prentice Hall, 1965).

occurrence of an act of terror in a given year (Yes or No) and measures of changes in income, poverty, employment, and manufacturing and agricultural output for the previous year.

_____

_____

_____

_____

b. A social scientist wonders whether the Sunbelt states are as politically conservative today as they are reputed to be. She believes that, apart from perhaps social issues, the opinions and beliefs of people in different regions are roughly the same. Moreover, she thinks that any variation among regions stems mainly from differences in the social class composition of the citizens living in those places. Her data are in the "American National Election Study 2000" data file that has categorical measures of region, attitudes on economic issues, and demographic characteristics such as income, education, and ethnicity.[7] What would be a good way to explore this hypothesis with these data?

_____

_____

_____

_____

_____

_____

_____

c. A Washington think tank wants to know why some states have more generous health care benefits for the poor than do others. It hypothesizes that two general factors explain the difference: states' overall political philosophy (degree of liberalism, for example) and economic capacity. The more liberal and wealthy a state, the more generous its health programs. The group's research firm has numerical indicators of health spending per capita for the poor, ideology,[8] percentage voting Democratic in national and state elections for the past ten years, per capita income, and economic growth over the past year. What method do you suggest?

_____

_____

[7] This topic would make a great research project for a student with access to the American National Election Study data and statistical software.

[8] Measures of states' ideology exist and are widely used. See, for example, Gerald C. Wright, Robert S. Erikson, and John P. McIver, "Measuring State Partisanship and Ideology with Survey Data," _Journal of Politics_ 47 (1985): 469–89. Also see William D. Berry, Evan J. Rinquist, Richard C. Fording, and Russell L. Hanson, "Measuring Citizen and Government Ideology in the American States, 1960–93," _American Journal of Political Science_ 42 (1998): 327–48.

_____

_____

_____

d. A student in a political communications class wants to know if a newspaper's size, as measured by average weekly circulation, affected its coverage of the 2004 National Democratic Convention, as indicated by the average number of column-inches it devoted to the subject in the month of July. He has spent considerable time collecting these data for forty papers throughout the United States. But he wonders whether coverage will also be influenced by the papers' political partisanship and region, for which he has only indicator or categorical variables. (Party bias, for instance, is coded into three categories—"pro-Democratic," "neutral," and "pro-Republican"—whereas region is indexed simply as 0 for "non-South" and 1 for "South.") Can you help this student pick an appropriate statistical strategy for investigating the hypothesis? How would your strategy handle the independent variables?

_____

_____

_____

_____

_____

_____

_____

Exercise 14–3. Multivariate analysis simply means the simultaneous examination of three or more variables. With categorical data, researchers *can* analyze contingency tables. Empirical social science has in practice, however, largely gone beyond this type of analysis, but the examination of multi-way contingency tables provides an excellent introduction to more advanced methods. The general idea is easily grasped. It consists of three steps:

1. Examine the relationship between the independent and dependent variables in a bivariate or two-way table. Ask yourself, How strong is the relationship? How are the individual categories related? (For example, you might find in a study of opinions on the environment that level of education is related to concerns about global warming: high school graduates are more concerned about global warming than are those individuals with less education by 22 percentage points.) Call this the *original relationship*. (Table 14–1 gives a new example.)

TABLE 14–1
Original Relationship: Voter by Gender

Vote in 2004	Male	Female
Kerry	46.4%	53.2%
Bush	53.6	46.8
Total	100%	100%
	(377)	(408)

Chi-square: 3.589; gamma: −.135; tau-*b*: −.068

Source: 2004 American National Election Study.

2. Analyze the two variables within each stratum of a control variable. That is, you will have a two-way table for each level of the control variable. Construct a multi-way table, as illustrated in table 14–2. Call this complex table the *controlled relationship.*

TABLE 14–2

Partial Relationships: Vote by Gender by Party

Vote	A. Democrats		B. Independents		C. Republicans	
	Male	**Female**	**Male**	**Female**	**Male**	**Female**
Kerry	89.9%	91.5%	55.6%	64.3%	8.5%	7.3%
Bush	10.1	8.5	44.4	35.7	91.5	92.7
Total	100%	100%	100%	100%	100%	100%
	(158)	(212)	(27)	(14)	(188)	(177)
	Chi-square: .290; gamma: −.097; tau-*b*: −.028		Chi-square: .290; gamma: −.180; tau-*b*: −.084		Chi-square: .169; gamma: .080; tau-*b*: .022	

3. Draw a conclusion. If the original relationship remains more or less the same in each level of the control variable, then the control variable does not have an impact; however, if the original relationship is weaker (or disappears) in most of the subtables, you have reason to suspect a spurious association. Or, if the relationship is not spurious, the control variable at least affects it statistically. It is also possible that the *Y-X* association will be even stronger in one or more of the subtables, a situation called *interaction.*

Here are some exercises to help you grasp the meaning of these ideas. The emphasis here is on conceptual understanding, not computation. Look at the two-way table showing vote for president in 2004 and gender (table 14–1).

a. Is there a relationship between vote and gender? Explain. (*Hint:* Pay attention to the percentages, the measures of association, and the chi-square).

_____

_____

_____

_____

b. The chi-square for this table is _____

What are the degrees of freedom? _____

Is there a statistically significant relationship at the .05 level? _____

We now introduce a control variable, party identification, which has three levels or categories: Democrat, independent, and Republican. Thus, there are three two-way tables, one for each category of party identification (table 14–2). Examine each table individually.

c. In subtable A (Democrats), is there a relationship between gender and vote? How strong is it? Is the chi-square significant at the .05 level? Explain.

_____

_____

_____

d. Give a similar analysis for subtable B (independents). Describe the relationship, if any.

_____

_____

_____

e. What is the relationship between vote and gender among Republicans (subtable C)?

_____

_____

_____

Exercise 14–4. A speaker at a lecture declares that no statistical evidence whatsoever demonstrates that the death penalty discriminates against minorities. When pressed on the point, she shows a slide that looks like table 14–3.[9] The subjects were 326 defendants indicted for murder in Florida during 1976 and 1977.

TABLE 14–3

Death Penalty?	Defendant's Race	
	White	Black
Yes	19	17
No	141	149

a. Do these data show a relationship between race and receiving the death penalty? Explain your answer in words and with a table and an appropriate test statistic.

_____

_____

_____

_____

_____

[9] These are "real" data first presented by M. Radlet, "Racial Characteristics and the Imposition of the Death Penalty," *American Sociological Review* 46 (1981): 918–27, cited in Alan Agresti, *Analysis of Ordinal Categorical Data* (New York: Wiley, 1984), 6.

_____

_____

_____

Now suppose another panelist says, "Wait a minute! These data tell only part of the story. If they are broken down by the race of the *victim* as well, we see that there is discrimination." He presents the data shown in table 14–4.[10]

TABLE 14–4

Death Penalty by Victim and Defendant's Race

	Victim's Race Z			
	White Defendant's Race (X)		Black Defendant's Race (X)	
Death Penalty? (Y)	White	Black	White	Black
Yes	19	11	0	6
No	132	52	9	97
Total	151	63	9	103

b. Can you interpret table 14–4? What, if anything, does it say about the discriminatory effects of the application of the death penalty in Florida during that time? (*Hint:* Look at each subtable one at a time and calculate percentages.)

_____

_____

_____

_____

_____

_____

_____

[10] Agresti, *Analysis of Ordinal Categorical Data*, 32.

INTERPRETING REGRESSION RESULTS

Here's a useful trick for understanding both the statistical and the substantive meaning of the coefficients of regression analysis. First, keep in mind the meaning of a multiple regression coefficient: it measures the amount Y changes for a one-unit change in a particular independent variable when all other independent variables have been held constant. This somewhat abstract definition can be made more meaningful when reading research results by following these steps.

1. Examine the summary table. For example, suppose, as is commonly the case, the findings are presented in a table like this:

Effects of Education and Race on Trust in the Judicial System

Variable	Coefficient
Constant	20***
Education (in years), X_1	2.0**
Race (0 for white, 1 for nonwhite), X_2	−3.0**

$R^2 = .45; ***p < .001, **p < .01.$

Here the dependent variable (Y) is a scale of trust in the judicial system; the higher the score, the greater the trust. There are two independent variables, education and race, with the latter coded as a dummy variable (0 for whites, 1 for nonwhites).

2. Write the regression coefficients, including the constant if there is one, as an equation. Place the numeric values of the coefficients in an equation:

$$\hat{Y} = 20 + 2.0\text{Educ} - 3.0\text{Race}.$$

Notice that we have included the constant term (20) and a minus sign before the coefficient for race. It is essential to keep track of the signs of the coefficients.

3. Imagine that the value of all independent variables is 0. Substitute this value into the equation. Example:

$$\hat{Y} = 20 + 2.0(0) - 3.0(0) = 20.$$

What does this mean? If a person had no schooling (0 years) and was white ($X_2 = 0$), the predicted level of trust ($\hat{Y}$) would be 20 units.

4. Now imagine that a person with no education ($X_1 = 0$) becomes nonwhite. Notice X_1 stays constant (that is, education is the same as in the previous equation), and only X_2 changes, from 0 to 1. Place these new values in the equation and simplify:

$$\hat{Y} = 20 + 2.0(0) - 3.0(1) = 20 - 3.0 = 17.$$

This result clearly demonstrates that race—when education is held constant—"causes" or is associated with a three-unit decrease in trust of the judicial system.

5. Now make another substitution. Go back to whites ($X_2 = 0$) and assume education increases by one year.

$$\hat{Y} = 20 + 2.0(1) - 3.0(0) = 20 + 2.0 = 22.$$

We see that an additional year of education increases trust among whites by 2 units.

6. Make additional substitutions until you have a feel for the impact of the variables. Suppose, for instance, a nonwhite individual has twelve years of education. What is her predicted trust score?

$$\hat{Y} = 20 + 2.0(12) - 3.0(1) = 20 + 24.0 - 3.0 = 41.$$

7. After making these kinds of changes, you should be able to write a substantive summary of the regression model. In this case it appears that as people acquire more education, they have more trust in the court system. This is true of both whites and nonwhites. But for a given level of education, trust among nonwhites is lower than among whites. Your next step would be to expand on this conclusion. You might hypothesize that both poorly educated people and members of minority groups have different experiences in the criminal justice system. Of course, that argument goes beyond the data presented here.

8. The estimated coefficients are the heart and soul of the research. The information at the bottom of the table—this format is typical of published articles—shows how well the data fit the linear regression model. The multiple regression coefficient, R^2, indicates that about half of the observed variation in trust scores is "explained" by education and race. The asterisks beside the coefficients indicate the level of significance. All three estimates are statistically significant, one at the .001 level, the others at the .01 level. (It is common for the coefficients in a model to have different degrees of significance.)

Exercise 14–5. Table 14–6 in the textbook presents the regression coefficients for the relationship between income inequality as measured by the GINI coefficient and two independent variables (union density and labor protections). In which of the following states would turnout be predicted to be highest? Why? (*Hint:* Use the least squares equation to predict income inequality.)

STATE 1: Union density = 1; Labor protections = .57

STATE 2: Union density = 2; Labor protections = .36

_____

_____

_____

_____

_____

_____

Exercise 14–6. Table 14–5 contains the results of a multiple regression analysis of the effects of development status (X_1), GNP (X_2), and infant mortality rates (deaths per 1,000 live births; X_3) on an indicator of political freedom, voice, and accountability (Y). The raw data can be found in chapter 11. The status variable is an indicator, or dummy variable, that is coded 0 for developed nations and 1 for developing nations.[11] The dependent variable measures "the extent to which a country's citizens are able to participate in selecting their government, as well as freedom of expression, freedom of association, and free media. The index is scaled so that the mean of all scores is 0 and the standard deviation is 1.0. The higher the score, the greater the freedom."[12] The observed range goes from –1.870 to 1.450 or 3.32.

TABLE 14–5

Voice and Accountability Regressed on Development Status, GNP, and Infant Mortality

Variable	Coefficient (standard error)	Mean	Observed Minimum	Observed Maximum
Constant	.9533 (.3415)			
Status	–1.4558 (.3270)	.5278	0	1
GNP	.00000798 (.00000925)	18450	600	68,800
Infant mortality	–.00145 (.002843)	29.44	2	118

[11] These codes differ from those presented in table 11–1 because we want to simplify the interpretation.
[12] Daniel Kaufmann, Aart Kraay, and Massimo Mastruzzi, "Aggregate and Individual Governance Indicators for 1996–2005," World Bank Policy Research Working Paper 4012, September 2006.

a. What is the estimated regression model?

_____

_____

_____

b. Interpret the coefficient for status (X_1). Do not provide a mechanical answer, but try to explain in terms an informed nonstatistician can understand.

_____

_____

_____

_____

c. The coefficients in the table are called *partial coefficients*. Explain this term to a layperson. Use the estimated coefficient -1.4558 in the explanation.

_____

_____

_____

_____

Exercise 14–7.   Political scientist Paul Goren investigated the relationship between core values and beliefs and policy preferences.[13] Goren defined *core beliefs* as "general descriptive beliefs about human nature and society in matters of public affairs," while *core values* are "evaluative standards citizens use to judge alternative social and political arrangements." He wanted to know if, how, and under what conditions these perceptions and attitudes affect opinions on public policy such as governmental welfare programs. Table 14–6 contains a small portion of his research results.

TABLE 14–6

Social Welfare Policy Opposition Regressed on Eight Independent Variables

Variable	Coefficient	(Standard Error)
Constant	11.99	(.98)
Race	−1.57	(.45)
Gender	−.76	(.30)
Family income quartile	.03	(.14)
Party identification	.43	(.14)
Feelings toward beneficiaries	−.03	(.00)
Economic individualism	.20	(.04)
Equal opportunity opposition	.58	(.06)
Political expertise	.03	(.07)

$R^2 = .48$; $F_{9,637}$; $df = 73.42$; $N = 638$

Source: Goren, "Core Principles and Policy Reasoning in Mass Publics," Table 2.

---

[13] Paul Goren, "Core Principles and Policy Reasoning in Mass Publics: A Test of Two Theories," *British Journal of Political Science* 31 (January 2001): 159–77. The coefficients used in this assignment differ from those in the published table. The changes were made after personal communication with the author.

The variables are as follows:

- Dependent variable, Y: Opposition to government social welfare provision. Additive, 25-point scale for which higher scores indicate greater opposition to governmental welfare services and spending
- Independent variables, X_k:
  - X_1: Race. 0 if white, 1 if African American.
  - X_2: Gender. 0 if male, 1 if female.
  - X_3: Income. Family income quartile (1–4).
  - X_4: Party identification. seven-point scale: 0 for strong Democrat, 1 for Democrat, 2 for independent-leaning Democrat, 3 for independent, 4 for independent-leaning Republican, 5 for Republican, and 6 for strong Republican.
  - X_5: Feelings toward beneficiaries. Summation of thermometer scores of feelings toward blacks, poor people, and people on welfare. The higher the score, the warmer the feelings.
  - X_6: Economic individualism. Scale of belief that "hard work pays off." Twenty-five-point additive scale with high scores indicating that being industrious, responsible, and self-reliant leads to economic success.
  - X_7: Equal opportunity opposition. Belief "that society should do what is necessary to ensure that everyone has the same chance to get ahead in life." Thirteen-point additive scale with higher scores indicating increased opposition to efforts to promote equality.
  - X_8: Political expertise. 0–8 scale of factual political knowledge, with higher scores indicating more information and "sophistication."[14]

a. What kind of variable is gender? _____

b. Write out the regression equation for this model.

_____

_____

c. Provide a short verbal interpretation of the partial regression coefficient for income (X_3). Assume you are explaining to someone who knows political science but not multiple regression.

_____

_____

_____

d. Give a substantive interpretation or explanation of the coefficient for gender (X_2).

_____

_____

_____

---

14. Ibid., 164–66.

e.   Consider only men (that is, assume $X_2 = 0$). If all of the other variables have scores of 0, what is the predicted value of the dependent variable?

f.   Can you offer a substantive interpretation of this value? That is, explain its meaning to a politician.

_____

_____

_____

_____

g.   Again consider just male respondents. What is the numerical effect of being African American if all other variables besides race have scores of 0? (Admittedly, this assumption doesn't make much sense—no one could be in the 0th income quartile, for example—but doing so makes it easier to interpret the individual coefficients.)

_____

h.   Consider an African American female (that is, $X_1 = X_2 = 1$) with an expertise score of $X_8 = 2.87$. Assume all the other variables are 0. What is the predicted opposition scale score?

_____

i.   The author of this study reports that the average score for economic individualism (X_6) is 13.70, and the mean for equal opportunity opposition (X_7) is 3.53.[15] Consider an African American female independent (that is, $X_4 = 3$) with an expertise score (X_8) of 2.57 and in the second quartile of income ($X_3 = 2$). Assume the score on feelings toward beneficiaries (X_5) = 50. If this person has average scores on individualism and equal opportunity opposition, what would be her predicted score on opposition to social welfare?

_____

j.   Suppose this same individual switches to strong Democrat from independent but retains the same characteristics (measures) on all the variables in the previous question. What is the effect of this change in party identification? _____ What is her predicted value on Y, social welfare policy opposition?

_____

k.   What does the observed F with 9 and 637 degrees of freedom tell you?

_____

_____

_____

l.   Given the N in the table, what is the appropriate sampling distribution to test the significance of individual coefficients?

_____

_____

[15] Ibid., table A, 176.

m. The standard errors for the coefficients appear in table 14–6. What is the observed test statistic for gender? Is it statistically significant at the .05 level? At the .01 level? (Use a two-tailed test for both tests.)

_____

n. What is the observed test statistic for political expertise (X_8)? Is it statistically significant at the .05 level? At the .01 level? (Use a one-tailed test.)

_____

o. What does the R-squared tell you about the fit of the data to the model?

_____

_____

Exercise 14–8. You and your best friend are working on a joint project about the role of religion in American politics. In particular, you want to know if people of different religious faiths disagree with respect to their evaluations of various political leaders and groups. You suggest using the 2002 American National Election Study because it has feeling thermometers for many public officials, interest groups, and parties. These variables can be treated as numeric, since the respondents can place themselves anywhere on a 100-degree (point) scale. (For examples, look at the "anes2000" files on the Web site http://psrm.cqpress.com.) But your friend objects that the main independent variable is nominal, and respondents are simply assigned to the categories "Protestant," "Catholic," "Jewish," "None," and "Other."[16] He's worried that you won't be able to take advantage of regression analysis, which the instructor wants you to use. Do you have an answer for this person's concern? What is it? How can you regress, say, feelings toward former President George W. Bush on religion? Be specific to demonstrate your knowledge of the method.

_____

_____

_____

_____

_____

Exercise 14–9. Your boss has asked you to critique a paper, "The Causes of Crime in Urban America." The authors have data on about seventy-five cities. Among many other analyses, there is a regression of Y, violent crime rate (offenses per 100,000 population in 1992), on X_1, per capita income in central cities, and X_2, per capita expenditures for public safety. (These data can be found in the files "stateofcities.dat" or "stateofcities. por" on the Web site http://psrm.cqpress.com.[17]) Luckily, some summary statistics for the variables have been supplied in an appendix to the report (table 14–7).

---

[16] The study has many codes for religion, but assume you want to use just those five.

[17] The data are from Norman J. Glickman, Michael Lahr, and Elvin Wyly, "The State of the Nation's Cities: Database and Machine Readable Documentation," version 2a (January 1998), Rutgers University: Center for Urban Policy Research. Available at http://policy .rutgers.edu/cupr/sonc/sonc.htm.

TABLE 14–7

Summary Statistics on Urban Crime

Variable	Valid Cases	Minimum	Maximum	Mean	Standard Deviation
Crime rate (Y_2)	69	141	3,859	1,573.00	88.158
Per capita income (X_1)	77	$9,258	$19,695	$13,679.56	$2,410.431
Per capita safety expenditures (X_2)	77	$101	$1,133	$280.43	$128.018

Source: Norman Glickman, Michael, Lahr, and Elvin Wyly, "The State of the Nation's Cities: Database and Machine Readable Documentation," Version 2a (January 1998), Center for Urban Policy Research, Rutgers University. Retrieved from http://policy.rutgers.edu/cupr/sonc/sonc.htm.

a. The paper reports that the partial regression coefficient of crime on per capita income ($\hat{\beta}_{YX_1}$) is –.0076. Does this number mean that economic well-being in a city is unrelated to crime? Explain.

b. The standard error of this partial regression coefficient is .041. What is the observed t?

c. Assuming that there is a constant and two independent variables in the equation and $N = 66$ for this model, would you reject the null hypothesis that the population partial regression coefficient of crime rates on income is 0 based on a two-tailed test at the .05 level? Why?

d. Now examine the second independent variable in the model, per capita city government spending on public safety (for example, police and fire protection). The partial regression coefficient of crime on this variable is 3.090. What does this result mean in substantive terms? More specifically, does it represent evidence of a cause-and-effect relationship? Why?

e. The standard error of the partial regression coefficient of Y on X_2 is .749. Knowing this, do you think the null hypothesis of no partial linear association between crime and public safety spending should be rejected? Why?

_____

_____

_____

_____

_____

f. You notice that the report concludes that expenditures for police have a greater impact on crime than a city's standard of living. To make the point, the authors mention that the expenditure-standardized regression coefficient is twice the size of the one for per capita income, but, strangely, they do not report the actual values. Can you calculate them? And more important, is there any reason at all to make that inference based on these data? (*Hint:* Look at table 14–7 for the statistics you need. Note, however, that the table contains some extra information. You need to select the right statistics. Refer to the section on standardization in chapter 13 of the textbook.)

_____

_____

_____

_____

_____

Exercise 14–10. Martin Gilens and Craig Hertzman raise a crucial point for judging the condition of American democracy. They studied news coverage of the 1996 Telecommunications Act, which among other things loosened restrictions on the number of television stations a media corporation could own. Some companies own both newspapers and local TV stations, whereas others do not. The passage of the law, the authors assert, meant that "on average the loosening of ownership caps in the 1996 Telecom Bill benefited media companies that already owned many television stations, and did not benefit (and may have hurt) companies that did not own TV stations."[18]

Some of their findings "strongly indicate a relationship between the financial interests of newspaper owners and the content of their papers' news coverage." This is not a matter of editorial content. Instead, the study's authors believe that what appears and *does not* appear in a paper's news sections may reflect the economic self-interests of the publisher and not inherent newsworthiness or the needs of the public. But they are aware that the connection between financial interest and news content may be "spurious, due to other characteristics of the newspapers."[19]

---

[18] Martin Gilens and Craig Hertzman, "Corporate Ownership and News Bias: Newspaper Coverage of the 1996 Telecommunications Act," *Journal of Politics* 62 (May 2000): 372.
[19] Ibid.

To check this possibility they conducted a multivariate analysis, a portion of which appears in table 14–8. The unit of analysis (27 in all) is "newspaper." The dependent variable is the proportion of a paper's coverage of the act that was "negative" (that is, discussed the possible adverse consequences of the act). One of the independent variables was total weekly circulation, which provides an indicator of a paper's revenues, which in turn determine the size of its "news hole" or space for news content. ("Papers with higher circulations and larger news holes might be expected to publish more information about the telecommunications legislation and might mention a higher percentage of negative consequences as a result."[20]) They also looked at presidential candidate endorsements as a proxy measure of the paper's political leanings and the percentage of revenue from broadcast television "to test the possibility that reporting on [the bill] was influenced less by the number of stations owned than by the parent company's economic dependence on TV revenue."[21] The investigators broke the main explanatory factor—ownership of television stations—into two dummy variables:

Substantial = 1 if company has 9 or more TV stations in 1995
0 if company owned none
Limited = 1 if company owned 2 to 5 TV stations in 1995
0 if company owned none[22]

TABLE 14–8

Corporate Ownership and News Bias

	Estimated Coefficient	Standard Error	t Statistic	Level of Significance (two-tailed)[a]
Substantial	−.39	.13		
Limited	−.23	.12		
Circulation	.04	.02		
Endorsement	.04	.05		
Percentage of revenue from TV	−.05	.35		

$R^2 = .35$; $N = 27$

[a] The researchers used a one-tailed test.

Assume that they have a random sample of newspapers. The table gives you part of the results from their regression analysis plus space to write some of your answers.

a. Write a model or equation for the predicted value of the proportion of negative coverage. (They didn't report a constant, so just ignore that coefficient.)

_____

_____

_____

_____

_____

[20] Ibid.
[21] Ibid.
[22] It is not clear why Gilens and Hertzman chose this particular way to measure ownership, but the interpretation of the results is straightforward if you keep in mind the definition of dummy variables. Gilens and Hertzman, "Corporate Ownership and News Bias," table 3, 382.

b. Interpret the *R*-squared.

_____

_____

_____

_____

_____

c. Interpret in statistical and substantive terms the coefficients for substantial and limited ownership. (*Hint*: Refer to the discussion of dummy variables in chapter 13 of the textbook. Try to understand them using the logic employed in that section.)

_____

_____

_____

_____

_____

_____

d. What are the observed *t* values? Write them in table 14–8.

e. If you wanted to test each coefficient for significance, using a two-tailed test, what would be the critical *t* at the .05 level? (It's hard to tell from the authors' explanation, so assume 22 degrees of freedom.)

_____

_____

Which of the coefficients is significant at that level? Indicate with a "Yes." Better still, give the level of significance.

_____

_____

Exercise 14–11. Suppose that for a comparative government class you want to study the effect of globalization on citizens' political beliefs and behavior. You decide to concentrate on Britain's membership in the European Union (EU), a transnational organization of European states that some Britons fear will infringe on British sovereignty. In particular, you want to know if positions on this issue affected party and candidate choices in the 2001 British general election. Table 14–9 *summarizes* the results of a cross-tabulation of responses to the question, "Overall, do you approve or disapprove of Britain's membership in the European Union?," and the party the respondents voted for in that election (Labour, Liberal Democrat, and Conservative). Your hypothesis is that those who support membership in the EU will vote Labour, whereas those against it will support the conservatives. (These numbers come from the "British Election Study 2001" and can be found in the files "bes2001.dat" or "bes2001.por" on the Web site http://psrm.cqpress.com.)

a. Try to replicate our findings in table 14–9. *Note:* We eliminated all the minor parties (that is, Greens, SNP, Plaid Cymru, and "other") as well as "none." Your table won't match ours exactly unless you do the same.[23] Why?

TABLE 14–9

Results of Cross-Tabulation

Chi-square	df	Tau-*b*	Gamma	*N*
160.516	8	.190	.274	1,918

---

b. More important, interpret the results in the table. Supply both a statistical *and* a substantive answer to the question, "Is there a relationship between opinions on the EU and the direction of the vote?"

_____

_____

_____

_____

_____

c. Party identification (partisanship) has been found to have a strong connection with voting: party loyalists usually back their parties in elections. So perhaps the relationship you saw in 14–11b (assuming that you think there is one) can be explained by the effects of party affiliation on those two variables. As the text indicates in chapter 13, one way to find out is to control for partisanship by creating subtables based on the categories of the control variable. In other words, the previous cross-tabulation from which the chi-square, tau-*b*, and gamma were determined was a 3 × 4 table. To hold party identification constant we generated three such 3 × 4 tables, one for each level of party identification. (*Reminder:* For simplicity's sake, we are considering only the three major British parties.) Table 14–10 shows the results when we examined the attitude by vote relationship within the three categories of the control variable, party identification.

TABLE 14–10

Results of Multivariate Analysis

Subtable	"Level" of Party Identification	Chi-Square	df	Tau-*b*	Gamma	*N*
1	Labour	16.992	8	−.04	−.11	919
2	Liberal Democrat	4.076	8	.03	.156	250
3	Conservative	47.347	8	.16	.41	566

Again, you might try replicating these findings by controlling for party identification (variable 01). In any event, try to interpret these results. Does the party variable explain or cancel the original relationship? (*Hint:* [1] Reread the section of chapter 13 on multivariate analysis of categorical variables. [2] Compare each of the

---

[23] We coded the vote variable this way: 1 = "Labour," 2 = "Liberal Democrat," and 3 = "Conservative." We did so because we wanted the scores to run from least to most conservative.

statistics in this table with the corresponding ones in the previous attitude by vote table. [3] Try calculating a [weighted] average of tau-*b* or gamma or both and then compare it to the measure in the two-way table.)

_____

_____

_____

_____

_____

Exercise 14–12. Chapter 1 of the textbook described a study that showed the deleterious effects of negative campaigning on voter turnout.[24] This important finding did not go unchallenged, however. In the spirit of replication that we discussed in Chapter 2, political scientists Martin Wattenberg and Craig Brians published an article that "directly contradict[s] their findings."[25] They rested their case partly on the analysis of two surveys, the American National Election Studies for 1992 and 1996. The main independent variables were indicators of whether or not respondents remember hearing or seeing negative and positive political advertisements and whether they made comments about these ads; that is, the variables were coded 1 if "Yes, comments were made" and 0 if "No." The dependent variable was a dichotomy: Did the respondent vote or not? The researchers hypothesized that if being aware of attack ads does adversely affect citizenship, there should be a negative correlation between commenting on attack ads and voting. If, by contrast, exposure to such ads had little effect on potential voters, the relationship would be nil.

Besides these variables, they also included many other independent factors that might affect the decision to vote. Table 14–11 presents a small portion of their results for the 1996 survey respondents.[26]

TABLE 14–11

Logistic Regression of Turnout on Advertising Recall and Other Variables

Variable	Partial Coefficient	Standard Error
Negative ad comment: 1 = yes, 0 = no	−.2005	.1792
Positive ad comment: 1 = yes, 0 = no	.2652	.2806
Newspaper political news index[a]	.0337	.0113
Age in years	.0295	.0054
Campaign interest: 1 = somewhat, 0 = otherwise	.3824	.1672
Campaign interest: 1 = very much, 0 = otherwise	2.0460	.3260
Gender: 1 = female, 0 = male	.3195	.1533
Time from interview to election (in days)	−.0053	.0043
Independent leaner:[b] 1 = independents who lean toward a party, 0 = otherwise	.8183	.2647
Weak partisan:[b] 1 = weak partisan, 0 = all others	.7279	.2543
Strong partisan:[b] 1 = strong partisan, 0 = all others	1.7830	.2935
Race: 1 = white, 0 = nonwhite	.1962	.2057
Constant	−4.4223	.4223

N = 1,373. Percentage of respondents correctly predicted 81%, based on these variables plus others not in this table.
[a] Coded on scale from 0 to 28 with 28 being highest interest.
[b] Pure independent treated as reference category.

[24] Stephen D. Ansolabehere, Shanto Iyengar, and Adam Simon, "Replicating Experiments Using Aggregate and Survey Data: The Case of Negative Advertising and Turnout," *American Political Science Review* 93 (December 1999): 901–10.
[25] Martin P. Wattenberg and Craig Leonard Brians, "Negative Campaign Advertising: Demobilizer or Mobilizer," *American Political Science Review* 93 (December 1999): 891.
[26] Ibid., table 3, 894. The Wattenberg and Brians model contains other control variables. We eliminated them to keep things simple.

a. The coefficients in the table constitute the terms of a logistic regression model. Write the model as an expression for the predicted probability of voting:

_____

b. Write the model as an expression for the estimated log odds (logit) of voting:

_____

c. If someone had null (zero) values on all of the variables, what would be his predicted probability of voting? (By the way, why do we write *his* here?)

_____

What is the substantive interpretation of this probability? Does it make any sense?

_____

_____

_____

_____

_____

d. Now suppose a person is fifty years old but has null values on all the independent variables. Before doing any calculations, look at the coefficient for age. Do you think this estimated probability would be higher than the previous one? Why?

_____

_____

_____

_____

e. What is the predicted probability of voting for this fifty-year-old man? _____

In words, what is the effect of age on the likelihood of voting when everything else is the same?

_____

_____

_____

f. Consider this same person. What are the estimated *log* odds that he will vote? _____

What are the estimated *odds* that the individual will go to the polls on election day? _____

Write a commonsense interpretation of this latter estimate:

_____

_____

_____

g. Consider a sixty-year-old white female who is a strongly partisan Democrat, mentioned both negative and positive commercials, has a score of 14 on the political news index, is very much interested in the campaign, and was interviewed five days before the election. What is the predicted probability that this person will vote? What are the odds of her voting? (*Hint:* List this woman's values on each of the variables. [For instance, if she is "very interested," how would she be coded on that and the other interest variable?] Then substitute them into the equation for the predicted probability.)

_____

_____

_____

_____

_____

h. Compare the person in 14–12g with an identical male. Who is more likely to vote? Why?

_____

_____

_____

_____

_____

_____

i. Which of the coefficients in the table are significant?

_____

_____

_____

_____

_____

j. In your view, are the authors correct in saying that exposure to negative political commercials does not depress turnout? (*Hint:* Think about the main independent variable. Then consider a "typical" person, as we did in some of the previous questions. Get the estimated probability or odds of voting for this individual both when he or she mentions negative ads [that is, when the score is 1] and when no negative commercials are mentioned. How much do the probabilities or odds change?)

_____

_____

_____

_____

_____

Exercise 14–13.   This is a tricky assignment, but it is one that resembles a lot of actual political and social research. The term *white flight* has been used to describe demographic change in urban America since at least the 1960s. Using the city data mentioned in exercise 14–9, try to build an explanatory model of the change in the white population in central cities from 1980 to 1990. (Two versions of the data are available on the Web site http://psrm.cqpress.com, "stateofcities.dat" and "stateofcities.por.") That is, treat "percentage change in white population from 1980 to 1990" as the dependent variable and attempt to explain variation in change with indicators from various categories of possible explanatory factors. For instance, you hypothesize that crime, poor housing, increased minority density, pollution and congestion, declining city services, and loss of job opportunities are at least associated with the changing demographic composition of cities. Report your results in a form acceptable to your instructor.

Exercise 14–14.   Instead of "white flight," identify a different variable that might be interesting to analyze.

_____

_____

Exercise 14–15.   The Web site http://psrm.cqpress.com contains several other files of various types. Pick an issue or a problem in political science or politics and see if there are data on the Web site that might be appropriate for analyzing your chosen topic. You can use the procedures described at the beginning of this chapter to organize your thinking, even if the data are strictly nominal or ordinal.

PREPARING AND ORGANIZING A MULTIVARIATE ANALYSIS

Analyzing more than two variables at a time can be a daunting chore, even for experienced data analysts. The secret, we believe, is the same as for any academic undertaking: think before acting. In the case of multivariate analysis, careful planning is of utmost importance. Hence, we offer a few suggestions to help you organize your research:

- As we discussed in previous chapters of the textbook, it is essential that you state a few working hypotheses. If you sit in front of a computer before organizing your thoughts, you will soon be drowned in printout. We guarantee it.
- If you are given a data set, pick a likely dependent variable—something that might be important to understand or explain. Then ask yourself which of the other variables in the file might be related to it. If you are starting from scratch, you have more leeway. But in any case, try to convert these ideas into substantive hypotheses. (Remember, a hypothesis is a tentative statement subject to verification. The result of the test is less important than starting with a meaningful proposition. Why? Because whether one accepts or rejects it, something of value has been learned. Testing trivial propositions [for example, poverty among children is correlated with poverty among families] doesn't advance our knowledge of anything.)
- Similarly, think carefully about what would be appropriate indicators of general explanatory factors. Suppose, for instance, you believe that high crime rates encourage people to leave cities for the suburbs or countryside. If you are trying to explain migration to the suburbs, one of your independent variables would be crime, which can be measured by, say, homicide rates or property lost to theft. Whatever the case, think of broad explanatory factors and empirical indicators of those factors.
- Sometimes the choice of variables is straightforward. Frequently, however, you may need your imagination to construct indicators. Suppose you theorize that *changes* in population density explain something but the data at your disposal contain only the actual populations and areas of cities for 1994 and 2004. You first need to compute a density for each year by dividing total population by area to obtain, say, persons per square mile. Then you could calculate another indicator, "percentage change in population density from 1994 to 2004."[a] Most software programs make these sorts of transformations easy.
- Remember that the data are empirical *indicators* of underlying theoretical concepts. You can't expect them to be perfectly or even strongly related to the dependent variable or to one another. In general, if you find a model that explains 40 to 50 percent of the statistical variation in Y, you will be doing well.
- We strongly urge you to analyze each variable individually, especially the dependent variable. Use the methods described in chapter 11 of the textbook.
- If your analysis involves multiple regression, first obtain plots of all the variables against each other. Doing so will reveal important aspects of the relationships, such as curvilinearity, the existence of outlying points, and the lack of variation in one or both variables. All these aspects can and should be taken into consideration.
- It is permissible, even advisable in some instances, to delete the outlying cases if you can do so on substantive grounds. Or, you can sometimes transform variables. Have you noticed, for instance, that some authors we cite in this book analyze not income but the *logarithm* of income. Doing so can mitigate the effects of a few very large numbers on a statistical procedure. One of the unfortunate properties of regression as we have introduced it is that it can be very sensitive to extreme scores. Hence, you may find that a statistical result changes considerably after adjusting the data.[b] *Just be sure to describe your methods fully in your report.*
- It helps to obtain a correlation matrix of your variables. This table will point to variables that are not related to much of anything and that might be dropped from the analysis. Equally important, correlation coefficients will help you decide whether an independent variable is related *in the way* your hypothesis predicts. If there should be a negative relationship, for instance, and the correlation is positive, your starting assumption may be wrong or you may have to look more carefully at the variable's definition.
- A correlation and a plot can also flag another possible problem. Sometimes two independent variables are so highly correlated that they are practically equivalent to each other, and including both in a regression model just adds redundancy; no separate information gets included. After all, if you regress height on weight measured in pounds *and* in kilograms, you just have two versions of one concept. Thus, if you spot high intercorrelations among the independent variables, ask if they are measuring the same thing or represent interesting substantive relationships.

Your final model may be much simpler than your initial expectations. That's probably a good thing because the goal of science is to find the simplest equation that has the highest predictive capacity. It is not important to include lots and lots of variables. One technique is to add or subtract one variable at a time and determine if it appreciably changes the model.[c] You may be able to eliminate quite a few variables this way, thereby reducing complexity.

---

[a] We recommend the log percent change, which is $\log^{10}$ (first number/second number) $\times$ 100, where $\log^{10}$ is the logarithm to the base 10.

[b] Trust us: far from "cooking the books," this maneuver is acceptable statistical practice in many instances.

[c] Lots of software programs have procedures for automating this process, but we don't recommend using them at this stage.